Using Basic Statistics In The Behavioral Sciences

2.362
.001/2

2.36
.01/2 = .005

2.36 ∓ .005
for upper & lower limits

4.126 — 5.621
.001/2 =.0005
.001/2 =.0005 } first this not 5.001
4.126-.005
5.621 +.005
4.1255 — 5.6215

Using Basic Statistics In The Behavioral Sciences

Annabel Evans Coldeway

Prentice-Hall Canada Inc., Scarborough, Ontario

Canadian Cataloguing in Publication Data

Coldeway, Annabel Evans, 1949–
Using basic statistics in the behavioral sciences

Includes index.
ISBN 0-13-943606-5

1. Statistics. 2. Psychology—Statistical methods.
I. Title.

QA276.12.C64 1989 519.5′02′415 C88-094565-6

Prentice-Hall, Inc., Englewood Cliffs, New Jersey
Prentice-Hall International, Inc., London
Prentice-Hall of Australia, Pty., Ltd., Sydney
Prentice-Hall of India Pvt., Ltd., New Delhi
Prentice-Hall of Japan, Inc., Tokyo
Prentice-Hall of Southeast Asia (Pte.) Ltd., Singapore
Editora Prentice-Hall do Brasil Ltda., Rio de Janeiro
Prentice-Hall Hispanoamericana, S.A., Mexico

ISBN 0-13-943606-5

Production Editors: Kirsten Hanson/Dick Hemingway
Design: Brian Lehen
Production Coordinator: Sandra Paige
Typesetting: Q Composition Inc.

1 2 3 4 5 JD 93 92 91 90 89

Printed and bound in Canada by John Deyell Company

To Peggy Runquist

Table of Contents

CHAPTER 3 DESCRIBING THE CENTRAL TENDENCY OF DISTRIBUTIONS

CHAPTER 4 DESCRIBING THE VARIABILITY OF DISTRIBUTIONS

CHAPTER 5 DESCRIBING THE POSITION OF SCORES IN DISTRIBUTIONS

CHAPTER 6 INTRODUCTION TO INFERENCE: THE NORMAL CURVE

CHAPTER 7 INTRODUCTION TO INFERENCE: PROBABILITY

CHAPTER 8 INTRODUCTION TO INFERENCE: THE RANDOM SAMPLING DISTRIBUTION

CHAPTER 9 INFERENCE WITH THE NORMAL CURVE

CHAPTER 10 INFERENCE WITH THE t-DISTRIBUTION

CHAPTER 11 INFERENCE WITH THE F-DISTRIBUTION

CHAPTER 12 ANALYSIS OF VARIANCE WITH REPEATED MEASURES

CHAPTER 13 MULTIPLE COMPARISON PROCEDURES

CHAPTER 14 INFERENCE WITH THE CHI-SQUARE DISTRIBUTION

CHAPTER 15 ADDITIONAL NON-PARAMETRIC TECHNIQUES

CHAPTER 16 PEARSON'S CORRELATIONAL TECHNIQUE

CHAPTER 17 PREDICTIVE TECHNIQUES

CHAPTER 18 TOOLBOX

APPENDIXES

Preface

Like most graduate students in psychology, I had to master graduate-level statistics in a course designed to separate the "women from the girls" and that we approached with a great deal of trepidation. Unlike most graduate students in psychology, I thoroughly enjoyed the course and learned a great deal. I was fortunate to have an instructor who obviously appreciated the elegance of statistical procedures, and his enthusiasm for the subject matter was contagious.

I began writing this book because I could never find a text that quite suited me. Many were too quantitative, with little space given to the underlying logic of statistics. Others used such complex notation and introduced so many arithmetic and algebraic derivations that my students were unable to see through the math and to get any conceptual grasp of the topic. So, I decided to write up my lecture notes and use them as my basic text. Well, as things are wont to do, that initial step escalated into a major project and this book is the result.

I have tried to keep the style of the book casual. Students are wary of statistics at the outset and I have never understood why authors of stat books promote these fears through dry, awkward and boring prose. I think that statistics are fun and I hope that this feeling comes through in my writing. A little humor can do a lot to relax an anxious reader.

Users of the book will notice the "Canadian content" in the examples. Of course "statistics are statistics are statistics" but, since I am a Canadian and my students are for the most part Canadian as well, I saw no reason to use the "number of slam dunks made by Magic Johnson" as my data base when I could use the "number of shots on goal by Glenn Anderson" instead. In a similar vein, I tried to give equal time to each gender in my examples to avoid the stereotypical emphasis on the male found in most textbooks.

I do not emphasize mechanical number crunching or rote memorization in my course, and the book reflects this orientation by avoiding unnecessary notation and jargon. I am not particularly interested in ending up with a student who can run a t-test or an Analysis of Variance. What I am interested in is ending up with a student who knows which is appropriate and why. With the technology available today a trained monkey can push a button to run a particular test of significance, but not even the most sophisticated computer can examine a set of data and decide how best to analyze them in order to answer the questions posed by the researcher.

The book is organized according to the format I use in my classes. I begin with descriptive techniques and follow these chapters with several designed to prepare students for inference. Inferential statistics are difficult for students to understand and I find it worthwhile to spend a lot of time laying the groundwork before I introduce inference.

The decision to place the chapters on correlation and regression at the end of the text was more or less arbitrary. Some other texts introduce these techniques early on in the course of study. I find the placement of these topics a matter of personal preference. There is no reason why these chapters could not be covered just before Chapter 9: Inference with the Normal Curve, since they can stand alone at that point.

The workbook that accompanies this text was prepared to parallel its sequencing and style. The problems are set up to be solved in a step-by-step fashion, much like the structure of text itself. A special section on "Choosing the Appropriate Test" allows students first-hand experience in making the decisions that researchers make. Although the chapter exercises provide some experience with this kind of decision-making, the workbook lays out the logical steps of the process and provides several additional examples and problems. All the solutions are available to the student in the back of the workbook.

The Instructor's Manual provides the "worked-out" solutions for the chapter and workbook exercises enabling instructors to see exactly where a student goes wrong. The test bank items at the end of the Instructor's Manual were prepared by one of my most successful statistics students. Lynn's diligence and conscientiousness has been truly impressive. Any errors that may occur in the test bank are no doubt due to my bungling.

If this book helps to alleviate anxiety, along with providing students with a basic understanding of statistical techniques, I will have accomplished my purpose.

ABC
Edmonton, 1989

Acknowledgments

I took my first undergraduate statistics course from Peggy Runquist, a professor in the Psychology Department at the University of Alberta. Although she was not a statistician by training, she had a "knack" for statistics. Her approach and style greatly influenced my own when it came time for me to teach a similar course. Peggy taught me that statistics can be a lot of fun to learn and to teach. I hope this book reflects that attitude.

A project like this is never accomplished without a lot of help from a lot of people. Susan Moisey did the initial edit of the manuscript. She replaced all my *which's* with *that's* and taught me the difference between *its* and *it's*. She also provided some of the more humorous examples in the book. Lynn Soetaert took on the mammoth job of preparing the test manual. Her work has been exemplary. Janis Watkin helped proof the galleys, a truly mind-numbing job. She changed a lot of Susan's *that's* back to *which's* and finally learned the difference between "μ" and "σ". Stan Rule, my statistics instructor in graduate school, helped reduce my confusion on several sticky points. David Walker gave me a lot of technical advice, such as how to convince a MacIntosh computer to put a bar on top of an "X"! Don Heth was kind enough to lend me his office and laser printer for a "short job" which turned into an entire day's work.

Several people reviewed the manuscript, and their comments and corrections were very helpful. They are G.K. Warriner, University of Waterloo; R.D. Seens, Concordia University; D.S. Lordahl, Brock University; G.E. Ebanks, University of Western Ontario, and V. Nallai-Nayagam, Mount Royal College.

Many people at Prentice-Hall have been involved in this project. Their commitment and support has been terrific. My copy editor, Dick Hemingway, has been particularly helpful and a great pleasure to work with.

My students at Concordia College "test-drove" the book through several versions and were diligent in finding errors.

I deeply appreciate the help all of these people provided. Thanks a lot!

ABC
Edmonton, Alberta

CHAPTER 1

Introduction to Statistical Concepts

Statistics as Tools in the Research Process

Statistics are tools for summarizing data, measuring relationships between sets of data, or making inferences about a large set of data by studying a subset drawn from it. This book attempts to show students in the biological, behavioral, and social sciences how to *apply* statistics to problems in their particular area of inquiry. The study of statistics per se is a mathematical discipline, much like geometry or algebra. This text, however, concentrates on how to *use* these techniques, provided by mathematicians, as tools to help us solve problems, in the same way that we use a slide rule or any other mathematical aid.

The Role of Statistics in Research

In research, statistics form the bridge from a question to a conclusion. Researchers begin with a question they wish to answer or some hypothesis they wish to verify. This question or hypothesis is then reworded in terms of data or evidence that might help to answer the question or verify the hypothesis. This is where statistics often come into play, helping the researcher to simplify or summarize data so that they can more easily be examined. Statistics may be used to evaluate data to determine if they adequately represent what the researcher wants to study.

The final step in research is to interpret the statistical analyses of the data and to make conclusions about the original question asked or hypothesis put forward.

Statistics, then, are tools the researcher can use to present or interpret information. There are various statistical techniques used by researchers. We will class these, for the purposes of this text, into three broad categories: **descriptive** statistics, **inferential** statistics, and **correlational/predictive** statistics.

Descriptive Statistics

Frequently, researchers need to gather large amounts of data regarding a variety of events. For example, we may wonder how well Canadian students do on high school departmental exams or how the salaries of Canadian physicians compare with those of their U.S. counterparts. When we gather information about questions such as these we are soon overwhelmed by numbers. It is very difficult to look at masses of numbers and see any trend or meaning in them. We need to summarize these numbers so that we and others can make sense out of them. Descriptive statistics serve this purpose. They allow us to *describe* a mass of numbers in terms of general trends, to tabulate data, and to present data in graphic form.

Using the tools of descriptive statistics, a mass of numbers can be presented in an organized and meaningful way, and data can be simplified so that their general trends can be seen.

Whenever our goal is to summarize, present, or organize a set of numbers for the purposes of clarification for ourselves and others, we are using the tools of **descriptive statistics.**

Inferential Statistics

Suppose that you are a psychologist interested in human memory. Let's say that you want to know how many numbers people can remember when you read off a list of 20 numbers. One way to answer this question would be to go out and test everyone and, if you did this, you would certainly have your answer. It would, however, take a lifetime or more for you to test everyone everywhere. Let's consider a less global question.

How do Catholics feel about the ordination of women? To answer this question you could send a questionnaire to every Catholic and ask, but again this would be incredibly expensive and time-consuming. A way around these

Population: The entire set of observations of interest.

Sample: A subset of a population.

dilemmas would be to find out how well *some* people remember or how *some* Catholics feel about the ordination of women, and then to generalize the findings to the whole population. For example, let's consider a water analyst who has come to test your drinking water.

She takes a *sample* from your tap, prepares a slide from that sample, examines it under a microscope and then gives you her assessment of the water. Although she has not examined all of the water, but just a small sample of it, she infers from that sample to all the water. Inference works like this. We study a small sample in some detail and then we generalize to the entire set of observations, called the population, from which that sample was taken.

Most psychological research is interested in people, all people. Psychological research, however, is not conducted on the population at large but rather on some portion of it. Inferential statistics provide tools for generalizing to the population at large by studying only a small sample of it. When used carefully and under appropriate conditions, researchers can use these tools to make *inferences* about populations by examining samples taken from them.

We use **inferential statistics** whenever we wish to infer things about the population at large, from information taken from a small sample of that population.

Correlational and Predictive Statistics

Everyday we notice relationships between things. For example, we all suspect that there is a relationship between studying time and grades. We know there is a relationship between diet and good health.

Researchers often need to describe relationships between things. The statistical tools that allow us to measure the strength of various relationships are called **correlational statistics.** Correlational statistics are descriptive when they describe the relationship between two entire sets of observations; they can also be used inferentially to infer from a sample the correlation in a total population.

A correlation often used by universities is the one between high school average and university performance. These two are related in that students with high grade 12 marks tend to do well in university. This kind of established relationship can be used in **predictive statistics**. If we know that school grades and college grades are related, then we can look at a student's school grades and *predict* how he or she will do in university. Predictive statistics provide tools for making predictions about an event based on available information. How well or how accurately we can predict, depends on how strong the relationship is between the two sets of observations.

When we are describing a relationship between events or when we are predicting from one event to another, we are using **correlational** or **predictive statistics**.

Preliminary Concepts

Constants versus Variables

A **constant** is just that—*constant!* A constant is a characteristic of objects, people, or events that does not vary. The temperature at which water boils (100 degrees celsius) is a constant.

A **variable** is a characteristic of objects, people, or events that can take on *different* values. It can vary in *quantity* (e.g., height of people), or in *quality* (e.g., hair color of people). Variables can be classified in different ways.

Classifying Variables

Experimental Classification

Independent variable: Variable controlled by the experimenter, and expected to have an effect on the behavior of the subjects.

Dependent variable: Some measure of the behavior of subjects and expected to be influenced by the independent variable.

Organismic variable: Characteristic of the subject, and not controlled by the experimenter.

A researcher may classify variables according to the function they serve in the experiment. An **independent variable** (IV) is one which the researcher usually controls. Its values are fixed by the experimenter.

For example, if I am interested in the effects of practice on problem solving, I might choose to give lots of practice problems to one group and only a few practice problems to a different group of people. After both groups have solved the practice problems, I could measure their performance on a new set of test problems. The independent variable is amount of practice and I have selected two values of that variable to use in my experiment.

A **dependent variable** (DV) is the one controlled (we hope) by the independent variable, not by the experimenter. The performance of my subjects on the test problems is my dependent variable. It is called dependent because we assume their test performance will *depend* on the independent variable (the amount of practice they received). Usually, in psychology research, the DV is some measure of performance, e.g., score on a test, preference measure, or reaction time.

There is a special class of variables that seem like independent variables but are not directly controlled by the experimenter. Age, gender and socioeconomic status, for example, are **organismic variables**. A great deal of research uses these kinds of variables. Technically they are not IV's because, although the experimenter can study them, he can not manipulate them. They are inherent in the organism and the researcher can only select groups of subjects who already possess certain levels of these variables.

Mathematical Classification

Continuous variable: Theoretically can take on an infinite number of values.

Discrete variable: Can take on a finite set of values.

Variables may also be classified in terms of the number of mathematical values they may take on within a given interval. **Continuous variables**, in theory, can take on any value within an infinite series of possible values. Weight, for example, is a continuous variable since the possible values for weight are infinite. Similarly, height is a continuous variable. In fact, the values of these continuous variables are limited only by the sensitivity of our scales or rulers, not by the variable itself.

Discrete variables can take on only a certain defined set of values. Number of children per family is a discrete variable in that certain values are not possible (e.g., 1/2 or 1 1/2). Gender is also a discrete variable; male and female are the only values possible.

Measurement Classification

Variables can be classified in terms of their scale of measurement. A **nominal variable** has values that differ only qualitatively. Eye color and gender are nominal variables. For simplicity,we may assign numbers to a nominal variable (e.g., male = 1, female = 2). These numbers, however, provide no

> **Nominal variable:** Has values which differ in quality only.
>
> **Ordinal variable:** Has values ordered by quantity.
>
> **Interval variable:** Has values ordered by quantity with equal sized intervals between each.
>
> **Ratio variable:** Like interval variables with a true zero point.

quantitative information. The numbers on the sweaters of hockey players are nominal. They are used merely for identification, not to indicate any measure of quantity. We wouldn't claim that a player numbered 99, for example, was 9 times better than one numbered 11.

Ordinal variables have values which are ordered according to quantity. Values are assigned to reflect that order. If we rated students and numbered them from best (1) to worst (50), we would have an ordinal scale. The intervals between the values, however, may not be equal. The second and third rated students are not necessarily the same distance apart as are the first and second rated students.

Interval variables have values where the intervals between them are equal. The Fahrenheit temperature scale is a good example of an interval variable. The difference between 40 degrees and 50 degrees Fahrenheit is the same as the difference between 60 and 70 degrees; that is 10 degrees. The interval variable has an arbitrary zero point. Zero degrees Fahrenheit does not mean zero temperature or the absence of heat; 80 degrees is not twice as hot as 40 degrees. IQ is another interval variable. A person with an IQ of 180 is not twice as smart as someone with an IQ of 90 because a zero IQ does not mean absence of intelligence.

Ratio variables are like interval variables but with a true or real zero point. Height, weight and the Kelvin temperature scale are examples of ratio variables. Distance is also a ratio variable. A field goal kicker whose maximum kick is 50 metres can kick twice as far as one whose maximum is 25 metres.

Introduction to Statistical Symbolism

Statistics has its own symbolic language and every statistician adds his or her own little idiosyncracies to the language of statistics. I am no exception.

In the beginning chapters of this text, pay particular attention to the symbols and abbreviations used. This will help avoid confusion later on.

Nominal - only differ in name, qualitative
Ordinal - quantitative, numbers diff
Interval - interval bet. #'s are equal
Ratio - scale starts at a true "0" point

Variables versus Constants

We will be using letters at the end of the alphabet to refer to variables, for example, X, Y, Z. Letters at the beginning of the alphabet are used to refer to constants, for example, a, b, c.

Notation

N: The total number of observations in a population distribution.
X_N: The last score in the X distribution
f: The number of times a particular score occurred in a distribution—stands for frequency
Σ: Sign for summation

The use of upper case letters at the end of the alphabet is reserved for naming collective sets or groups of scores. We refer to the X distribution or the Y distribution of numbers, for example. Sometimes X has a subscript $_i$. X_i refers to a particular score in the X distribution; X_1 is the first score in the X distribution, X_7 is the seventh score. Do not use lower case x or italicized *x* when discussing a group of scores.

N is the total number of scores in a distribution. If my psychology class has 42 students, then N = 42.

X_N is the last score in a distribution. On the final exam in my psychology class, $X_N = X_{42}$. This refers to the particular score obtained by the 42nd student.

f refers to frequency, usually the frequency of a particular score in a distribution. If 13 students got 63% on the final exam in my psychology class, then the f for 63% is 13. Do not use upper case F when referring to frequency. F has a different meaning entirely.

Σ is a summation sign. ΣX instructs you to sum or add up all the scores in the X distribution.

Statisticians use a particular kind of notation when they are referring to a population of scores and a different notation when they are referring to a sample drawn from a population. Descriptive statistical techniques are used to describe entire populations and so population notation should be used. An *entire* population does not mean a huge population; rather it means the population of interest. Inferential statistics are used to make inferences from a sample to a population and so sample notation should be used. Although we won't be using sample notation until the chapters on inference, many text books introduce sample notation early. The table below presents some of the standard notation used by statisticians to refer to populations and samples. We will be using these in later chapters.

Sample versus Population Notation

Measure	Sample	Population
Size	n	N
Mean	$\overline{X}$ or $\overline{Y}$	μ_x or μ_y
Standard Deviation	S or S.D.	σ
Variance	S^2 or Var.	σ^2

SUMMARY OF TERMS AND CONCEPTS

Descriptive statistics describe the characteristics of an entire set of data, called a population.

Inferential statistics use sample data to make inferences about populations.

Relationship and **predictive statistics** describe the relationship between X and Y or predict Y from X.

Variables take on more than one value but **constants** take on only one value.

Independent variables are **experimental variables** used by the experimenter to cause changes in **dependent variables**.

Continuous variables may have an infinite number of values but **discrete variables** have a finite set of values.

Nominal variables have values which differ in name only, whereas **ordinal variables** have values that are ordered with respect to quantity.

Interval variables have equal intervals between values and an arbitrary zero point.

Ratio variables are like interval variables with real zero points.

X refers to a set of scores as in the X distribution.

N is the total number of scores in a distribution.

f refers to the frequency of a score in a distribution.

Σ is a summation sign. It is an instruction to sum a set of scores as in ΣX.

EXERCISES

1. Label each of the following as discrete or continuous variables.

a) academic aptitude
b) number of students in psychology courses
c) reaction time
d) temperature
e) socioeconomic class
f) academic degree
g) age

2. A researcher wants to study the effects of three different brands of pain reliever on perceived pain relief. He gives one of the three brands to different groups of subjects suffering from chronic headache. He asks all subjects to rate the effectiveness of the drug on their headaches. What is the independent variable and what is the dependent variable?

3. Answer the following questions about the distribution of scores below.

Data: 23, 10, 9, 8, 8, 8, 8, 6, 4, 4, 2

a) N = ?
b) f for the score value 8 is?
c) X_8 = ?
d) X_N = ?

4. Label the following as nominal, ordinal, interval or ratio **and** as continuous or discrete.

a) Grading system using A, B, C, F ___ ___
b) Number correct on a quiz ___ ___
c) Fur color of dogs ___ ___
d) Numbers on football jerseys ___ ___

5. Part of my PhD dissertation involved comparing the time it took subjects to learn lists of words. The lists varied in terms of the relationships between the words in the list. For example, some of the lists were composed of words which were similar in meaning. Others had words that rhymed. What were my independent and dependent variables?

CHAPTER 2

Organizing and Presenting Data

In research, we frequently find ourselves with a great many numbers about some event. Only rarely can we simply "scan" these raw data and make any sense out of them. We must organize our numbers into some form that is consistent and understandable not only to us, as data collectors, but also to anyone else who might be examining the data that we have collected.

Organizing Raw Data

Imagine a researcher who wishes to compare two different high schools in terms of grade 12 student performance on a standardized achievement test. Here are the data.

				97	96	
					94	78
90			76			
	82	75				
		79	70		68	
	69					
65	66					
		64	58			
	63	55				
					62	46

As you can see these data are not easily interpretable!

Options are available to the data collector for organizing these scores. He should, of course, organize first by school. Within each school he can organize by value of the variable. In other words, he can list all the scores, from highest to lowest. Alternatively, he can organize by subject name, or number if he has assigned numbers to his test participants. As long as his technique is consistent, either method is fine for the initial organization of his data.

Organizing by Value

Holly H.S.	Rally H.S.
97	96
94	78
90	76
82	75
79	70
69	68
65	66
64	58
63	55
62	46

Organizing by Subject Number

Subject No.	Rally H.S.	Subject No.	Holly H.S.
1	46	1	82
2	75	2	64
3	66	3	90
4	78	4	69
5	76	5	94
6	55	6	62
7	68	7	79
8	70	8	63
9	58	9	97
10	96	10	65

Although the above example has only 20 data points, this initial organization is hard to interpret. We cannot, at a glance, tell which school seems superior. Imagine the chaos if there were hundreds of scores. Clearly, we need ways to organize data to present our findings.

Presenting Raw Data

Often researchers are called upon to report or present their data in journals or at conferences. Psychologists, as well as researchers in other disciplines, have several conventions that they follow for uniformity. This makes it easier for readers of reports and for convention audiences. Raw data that are organized for the purpose of presentation are ordered by values of a variable and by group. We do not present data by subject number in most cases.[1] There are two ways of presenting raw data, by table and by graph.

Presenting Data in a Table: The Frequency Distribution

A frequency distribution is a table indicating the values that a variable can take on and the frequency with which each value occurs.

Types of Frequency Distributions

There are two basic types of frequency distributions. A **univariate** frequency distribution has one variable; each subject contributes one observation. A **bivariate** frequency distribution has two variables; each subject contributes two observations.

Univariate Frequency Distributions

Score on Exam	Frequency (f)
90	2
89	0
88	4
87	3
86	1

Notice that two students received scores of 90 and no one received an 89 on the exam.

[1] Data are often presented by subject number when the researcher has used a single-subject design rather than a groups design. This is usually the case in the field of psychology called Experimental Analysis of Behavior.

Bivariate Frequency Distributions

		Score Value Variable One			
		71	72	73	74
Score Value Variable Two	9	1	2	0	0
	8	1	0	3	0
	7	3	0	2	2
	6	0	1	4	3
	5	4	4	1	3

Notice that one subject scored 9 on Variable Two and 71 on Variable One, whereas no one got a 71 and a 6.

Forms of Frequency Distributions

There are several different forms typically used to present data in a frequency distribution. The particular form a researcher selects will depend on the type of measurement, the number of values, and the purpose for presenting the data. We will look at two basic forms.

Simple Frequency Distributions To construct a simple frequency distribution, we list all possible values of the variable in a column and indicate the frequency of each value in a second column to the right. All possible values of the variable must be listed, including those that may not have actually occurred. If the variable is nominal the values may be listed in any order. With ordinal, interval and ratio data the values are listed from highest to lowest. All tables should be labeled with a number and a title and this label placed above the table. These are simply conventions that we follow for uniformity. Table 1 is an example of a simple frequency distribution of a hypothetical statistics class for the 1986 semester.

TABLE 1 *Simple Frequency Distribution of Statistics Grades*

Grade	f
9	2
8	0
7	4
6	7
5	8
4	6
3	2

Absolute frequency: The number of times a score occurs in a distribution.

Relative frequency: The number of times a score occurs in a distribution, divided by the total number of scores.

Cumulative frequency: Summing frequencies from the bottom of a frequency distribution up for each value or interval.

Frequently, we want to compare groups. Perhaps I would like to compare the performance of my class with a class taught by a different instructor. If the two classes are of different sizes, **relative** frequency rather than **absolute** frequency, is preferable. If, for example, I have a class of 50 students that I wish to compare with another class of 150 students, it is difficult to do so using absolute frequency. To say that 10 of my students received 9's compared to 20 students in my colleague's class would not tell us very much because of the huge difference in class size. In such a case it is much easier to interpret relative or proportionate frequency. For example, it means much more when I say that 20% of my class received 9's but only 13% of my colleagues' students did so. Relative frequency is extremely useful when we are comparing groups of different sizes. Relative frequency is also preferable when we are describing the performance of a single group of an unusual size. Reporting that 43% of 147 people think that Day Care should be heavily subsidized by the government is more meaningful than reporting the absolute frequency of 63. Always keep in mind that the purpose of presenting data is to make it easy for your reader or audience to see the trends in your results. The following table shows how we would present my statistics grades using relative frequency reported in decimal form.

TABLE 2 *Simple Frequency Distribution of Statistics Grades*

Grade	Rel f
9	0.07
8	0.00
7	0.14
6	0.24
5	0.27
4	0.21
3	0.07
Total	1.00[2]

There is a third way to report frequency. **Cumulative frequency** is used when we are interested in determining relative standing at a glance.

[2] This column must add up to 1.00

Cumulative frequency distributions add the frequency of each value to the total frequency below. Absolute or relative frequency may be used. Table 3 shows my statistics grades reported using absolute cumulative frequency.

TABLE 3 *Simple Frequency Distribution of Statistics Grades*

Grade	cf[3]
9	29
8	27
7	27
6	23
5	16
4	8
3	2

A quick glance at this distribution tells us that 23 of my 29 statistics students received grades of 6 or less. As you can see, this type of frequency is useful for determining relative standing. A student who received a 6 in the course can see that she performed better than at least 16 of the students.

Reporting the cumulative frequency in relative frequency form produces a relative cumulative frequency distribution. Again we would translate to relative frequency if we were comparing groups or if our group size was unusual. With a cumulative relative frequency distribution we are able, at a glance, to determine what proportion of the group obtained a particular score or below that score. Table 4 presents the grades of my statistics class using cumulative relative frequency.

TABLE 4 *Simple Frequency Distribution of Statistics Grades*

Grade	Rel. cf[4]
9	1.00
8	0.93
7	0.93
6	0.79
5	0.55
4	0.28
3	0.07

A student who received an 8 in this course sees, at a glance, that he performed as well or better than 93% of the class. This type of distribution, then, is helpful for quickly determining the location of a particular score in a distribution.

[3] The top value in the cumulative frequency column must equal the group size.

[4] The top value in any cumulative relative frequency distribution must always be 1.00 since the entire group must have received the top score or below.

Mutually exclusive interval: Non-overlapping interval in a grouped frequency distribution such that each score belongs in one interval only.

Width: The range of each interval in a grouped frequency distribution.

Mid-point: The middle value of an interval in a grouped frequency distribution.

Grouped Frequency Distributions Any simple frequency distribution lists all possible values of the variable, even those that did not occur. Often there are too many possible values for a simple frequency distribution to be practical. For example, a typical variable used in schools is percentage. If one student got 0 and another got 100 on a test, a simple frequency distribution of these data would require listing all 101 values in a column. In situations like these we often prefer to group data for the sake of clarity. As a rule of thumb, whenever there are more than 20 values of the variable, or when there are a large number of zero values in our data, a grouped frequency distribution is preferable.

To construct a **grouped** frequency distribution, the values are grouped into equal-sized intervals. These intervals are then listed from highest to lowest. Intervals must be **mutually exclusive**; each score belongs in one interval only.

When selecting an interval **width**, it is helpful to use an odd-sized rather than an even-sized one. This makes finding **mid-points**, which we will do shortly, much easier.

The following data (hypothetical) are the number of goals scored by 20 hockey players in the Canadian league in a single season. Let's use these data to construct a grouped frequency distribution. We will do this in a step-by-step manner.

DATA: 64, 63, 57, 56, 55, 54, 47, 46, 45, 45, 45, 44, 43, 37, 36, 34, 34, 23, 23, 15

Step 1. Determine how many *possible* values exist. The span of our data is 50 values (i.e., from 15 to 64). There are too many values for a simple frequency distribution to be practical. We would have a column of 50 numbers if we chose to do this.

Step 2. Decide how many intervals to use. It is conventional to use between 10 and 20 intervals. We will use 10.

Step 3. Determine the interval width. The symbol for width is **i**. We have already decided that we want to have 10 intervals, so we must determine what width will give us 10 intervals. Since we have a range of 50 values, we can see that an interval width of 5 will produce what we want.

Step 4. List the intervals from highest to lowest in a column and the frequencies in a second column.

TABLE 5 *Grouped Frequency Distribution of Hockey Goals*

No. of Goals	Abs. f	Rel. f	Cum. f	Rel. cf
60-64	2	0.10	20	1.00
55-59	3	0.15	18	0.90
50-54	1	0.05	15	0.75
45-49	5	0.25	14	0.70
40-44	2	0.10	9	0.45
35-39	2	0.10	7	0.35
30-34	2	0.10	5	0.25
25-29	0	0.00	3	0.15
20-24	2	0.10	3	0.15
15-19	1	0.05	1	0.05

If we look at the cumulative frequency and relative cumulative frequency columns, we can see at a glance that 14 hockey players of 70% of the hockey players scored 49 goals or less. These frequency distributions, then, are useful when we want to determine relative standing or location of a particular score in a distribution.

Presenting Data in a Graph

Frequency distributions can also be presented in graphic form. Many people find it easier to interpret data presented in a graph rather than a table.

Graphing Univariate Frequency Distributions

There are several conventions in graphing. These provide consistency in presentation.

1. Graphs are called figures.
2. The figure number and label are found below the graph.
3. Frequency is indicated on the ordinate (Y axis) and the values of the variable are on the abscissa (X axis).
4. A 3/4 rule is used generally (the ordinate should 3/4 as long as the abscissa).
5. Break any axes which do not begin at zero.

There are several types of graphs. The one you select will depend upon your purpose and the type of data you have. Absolute, relative or cumulative frequency may be used.

Bar Graph The **bar graph** is used to present discrete data. Separate bars represent each value of the variable and the height corresponds to the frequency with which each value occurred. The bars are separated to indicate that the variable is discrete rather than continuous. Nominal and ordinal data are often presented in a bar graph.

Suppose that I am thinking of going into the dog breeding business. I decide to collect data about the popularity of some breeds of dogs. These data will will help me decide what breeds to specialize in. I survey a large neighbourhood and determine the kinds of dogs people own in it. These data are nominal and should be depicted in a bar graph.

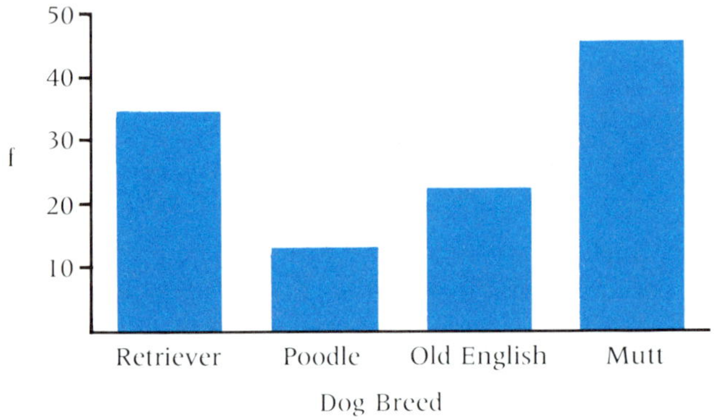

FIG. 1. *Popularity of dog breed*

The bar graph can be used to depict ordinal data as well as nominal data. The values of the variable should reflect their natural order. A hypothetical survey of student attitude toward the legalization of marijuana provided the data for the graph below. As you can see, relative frequency is on the ordinate or Y axis.

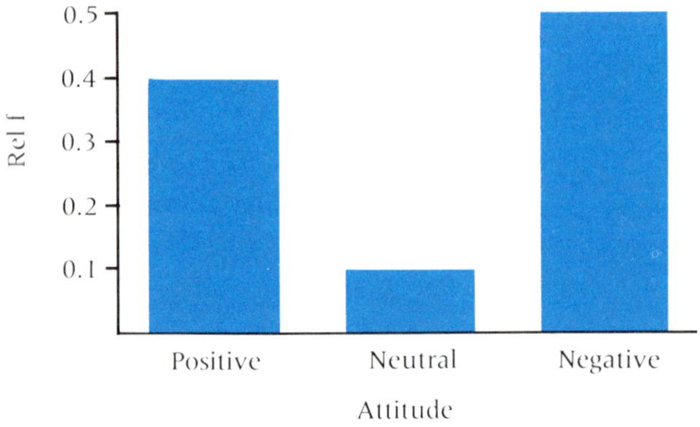

FIG. 2. *Attitude toward legalizing marijuana*

Exact limits: The mathematically precise beginning and end points of a score or interval. Also called real limits.

Histogram The **histogram** is similar to the bar graph except the bars are attached to indicate that the data are continuous. To construct a histogram we need to learn a new term: **exact limits**. When data are continuous, exact limits are the precise beginning and end points of a score or an interval. To obtain the upper exact limit we add half of the smallest unit of measurement to the score or upper limit of the interval. To obtain the lower exact limit we subtract half of the smallest unit of measurement from the score or lower limit of the interval.

For example, consider a bathroom scale which weighs to the nearest kg. The smallest unit of measurement, then, is 1 kg. I weigh 55 kg. The exact limits for my weight are 54.5 kg and 55.5 kg. If this procedure seems obscure, remember that my bathroom scale measures to the nearest kg. If, in reality, my weight was 54.62 kg, my scale would report my weight at 55 kg. If I weighed 55.43 kg, my scale would report me at 55 kg. When my scale says I weigh 55 kg, I may, in reality, weigh anything between 54.5 kg and 55.5 kg.

Here are some other examples.

Variable	Smallest Unit	Amount to + and −	Score or Interval	Exact Limits
Weight	1 kg	0.50 kg	56 kg	55.5–56.5 kg
Weight	0.5 kg	0.25 kg	50–53 kg	49.75–53.25 kg
Time	0.10 sec.	0.05 sec.	15.5 sec.	15.45–15.55 sec.

When constructing a histogram, the outsides of the bars are the exact limits and the middle of the bar is over the middle (**mid-point**) of the interval. A histogram may report the exact limits *or* the mid-point on the abscissa. I prefer to use mid-points because I think the graph looks better. Figure 3 presents student performance on a statistics exam.

Notice that the abscissa is broken to indicate that the test score variable does not begin at zero. The histogram is very useful graphing technique for continuous data. Most people have no difficulty interpreting data presented in a histogram.

Frequency Polygon Data presented in a histogram may also be presented in a **frequency polygon**. Points are used instead of bars. When data are

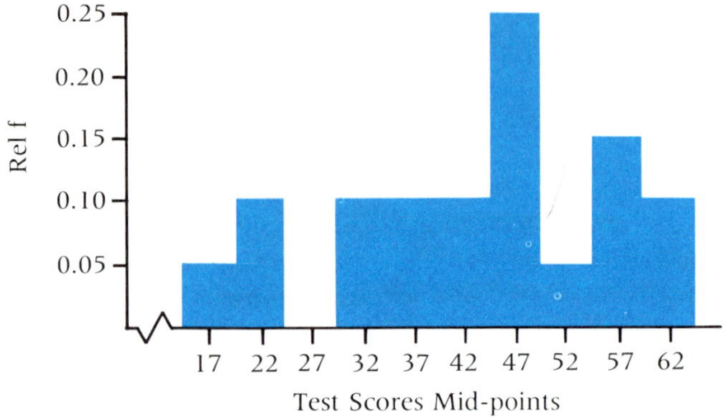

FIG. 3. *Student performance*

grouped, each data point is plotted over the mid-point of the interval; when data are in a simple frequency distribution, each point is plotted over the score value. The points are then connected by straight lines. Let's present the student performance data of Figure 3 in a frequency polygon.

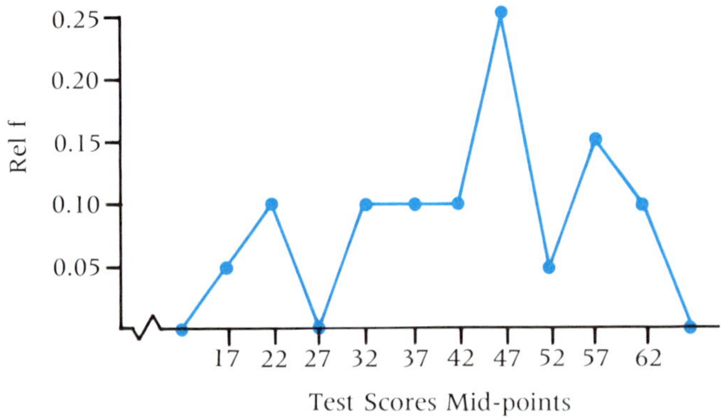

FIG. 4. *Student performance*

It is customary to drop the line to the abscissa at the mid-point of the adjacent interval at both extreme ends. In this way, the polygon is formed.

Ogive I have seen the **ogive** called a *cumulative frequency polygon* which I find very strange, since it is not a polygon at all. This graph is constructed from a cumulative frequency distribution. It is appropriate for continuous data and is often used when we wish to locate score position. Questions like "How have I done compared to the rest of the group?" and "What score would I have needed to be in the top 10% of the group?" can be answered by inspecting the ogive. Cumulative frequencies are plotted as points directly over the *upper exact limit* of each score or interval and the points are connected by straight lines. The ogive below presents our data in cumulative frequency.

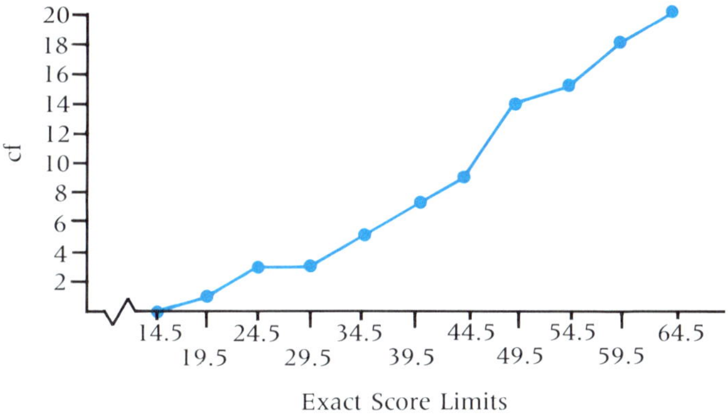

FIG. 5. *Student performance*

The ogive is useful for determining relative standing. A student who scored 49 on this exam can see, at a glance, that he performed as well or better than 14 of the 20 students in the class.

Graphing Bivariate Frequency Distributions: The Scattergaph

The graph used for bivariate frequency distributions is called a **scattergraph**. Recall that a bivariate frequency distribution has frequencies for two variables. Graphing these data can show us, at a glance, whether or not the variables are related or correlated.

The values of the first variable are located on the abscissa and the values of the second on the ordinate. Each subject contributes one point to the graph. The points are not connected.

As an example, imagine that I have collected sense of humor ratings and popularity ratings (provided by impartial judges) on 10 of my closest friends. I want to find out whether these two variables are related. In other words, are my friends who are rated the most popular also rated as having a good sense of humor? The data might look something like this:

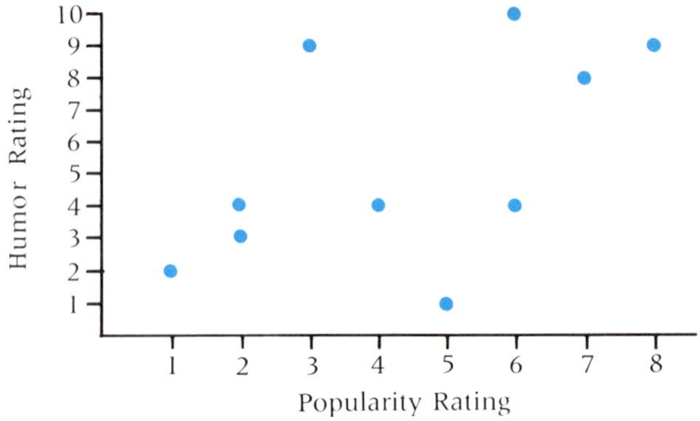

FIG. 6. *Ratings of popularity and sense of humour*

Each point represents the rating a friend got on the two variables. It looks as if there might be some relationship. Except for two individuals, those who are most popular seem also to have a good sense of humor.

A quick inspection of a scattergraph will give some indication of the degree of relationship between variables. There is a statistical technique which quantifies this relationship and we will be studying this in the chapters on correlation.

The Shape of Univariate Frequency Distributions

Univariate frequency distributions have two important shape characteristics: **symmetry** and **kurtosis**.

Symmetry

Positively skewed distribution: Graphically presented, a distribution with a tail pointing to the right of the graph.

Negatively skewed distribution: Graphically presented, a distribution with a tail pointing to the left of the graph.

Symmetrical distribution: One which is a mirror image about its center.

A **symmetrical** distribution is a mirror image about its centre. Some examples of symmetrical distributions used in statistics are the **normal** distribution and the **t**-distribution. Rectangular and U-shaped distributions may also be symmetrical. This is what they look like graphically.

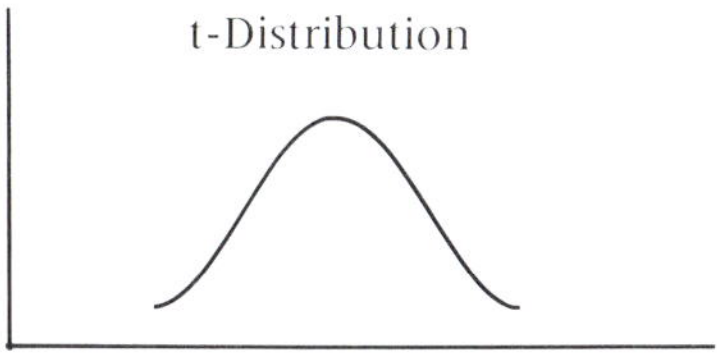

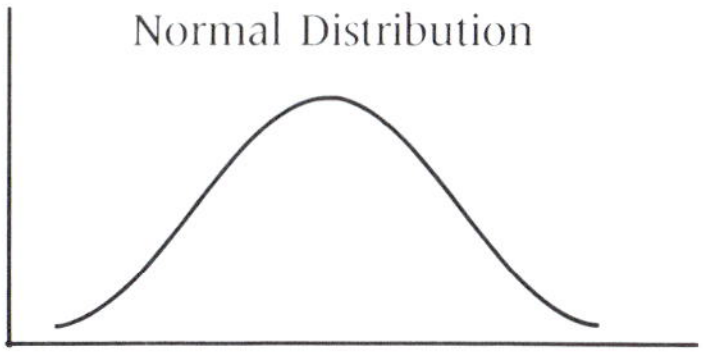

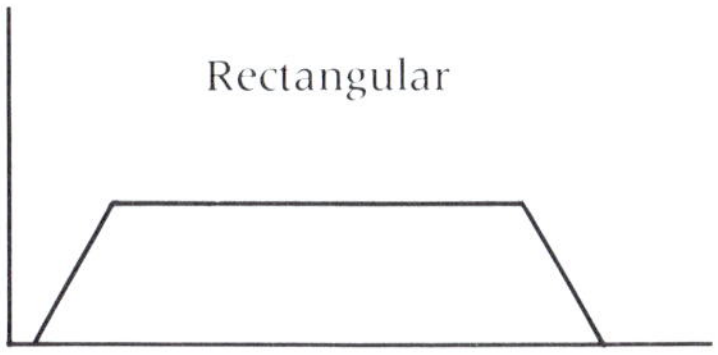

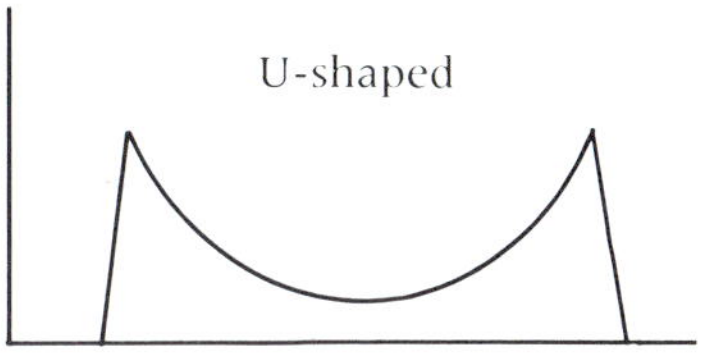

In a **skewed** distribution, the bulk of the observations (frequency) lies to one side. One side of the distribution is not a mirror-image of the other. Skewed distributions are described as having a positive or negative skew. When the extreme end or tail of the distribution points toward the positive side (the right) of the graph, we call this a **positive skew**. When the tail points to the negative or left side of the graph, we call this a **negative skew**. Examples used in statistics are the **chi-square** and the **F**-Distribution, both of which are positively skewed.

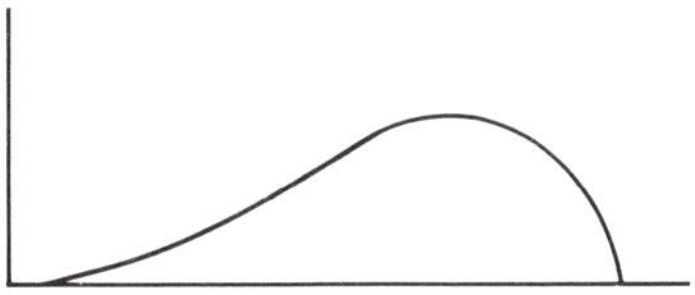

Negative Skew

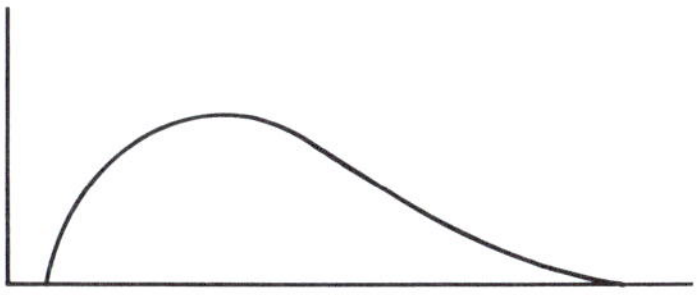

Positive Skew

Kurtosis

Platykurtic: Flat compared to the normal distribution.

Leptokurtic: More peaked with more area in the tails than the normal distribution.

Kurtosis is a shape characteristic describing the spread or scatter of the observations or scores. If each score has a similar frequency (i.e., occurred the same number of times), the distribution will look flat when it is graphed. Flat distributions are called **platykurtic** (think of the bill of a platypus). Distributions with lots of scores bunched in the middle are called **leptokurtic**.

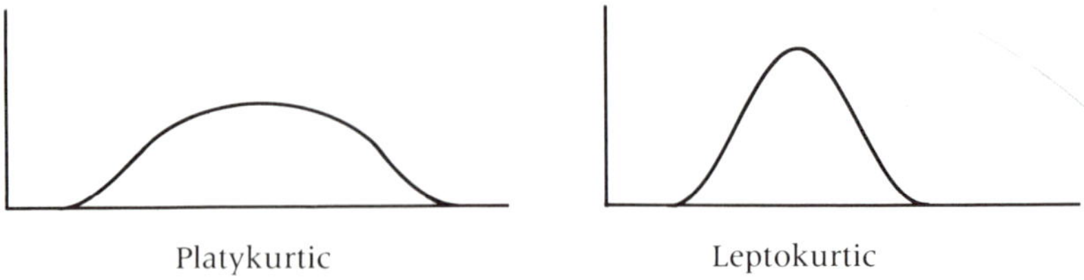

Kurtosis is defined in relation to the normal distribution, which will be discussed later. Basically, a distribution which is flatter or more variable than the normal is platykurtic, and distributions which are less variable, with more area in the tails than the normal, are called leptokurtic.

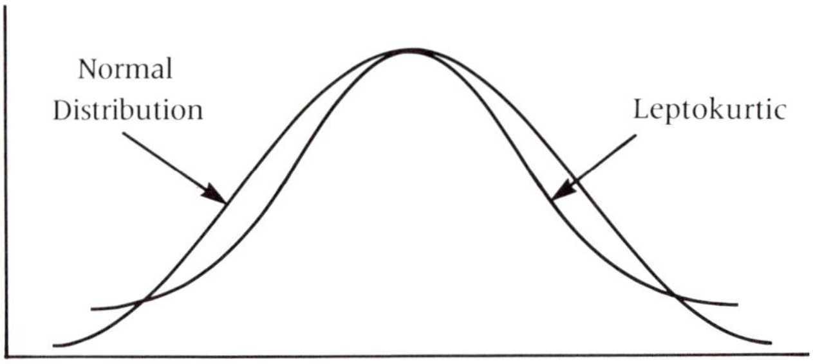

SUMMARY OF TERMS AND CONCEPTS

Raw data are initially organized by **subject number** or from the highest to the lowest value of the **variable**.

Data are often presented in tables, called **univariate** and **bivariate frequency distributions.**

A **simple frequency distribution** lists all possible values of the variable in a column accompanied by the frequencies with which each value occurred.

A **grouped frequency distribution** is used for data which have been grouped into equal sized intervals. The intervals are listed in a column accompanied by the total frequency of values within each. **Absolute, relative,** and **cumulative frequency** may be used.

Discrete variables are presented graphically in a **bar graph**. The separation of the bars indicates the discreteness of the data.

Continuous variables may be presented in a **histogram** or a **frequency polygon**.

The **ogive** is used to present cumulative frequency data and is useful for locating scores, graphically.

Bivariate data are presented in a **scattergraph**. The scattergraph provides a visual display of the relationship between the variables.

Symmetrical distributions have a left side which is a mirror image of the right.

Positively skewed distributions have the bulk of the frequency on the left side; **negatively skewed distributions** have the bulk of the frequency on the right side.

Distributions that are flatter than the normal distribution are **platykurtic**; distributions that are more peaked with more area in the tails than the normal are **leptokurtic**.

EXERCISES

1. Construct a simple frequency distribution for the following data.
DATA: 15, 12, 10, 9, 8, 18, 25, 11, 13, 14, 7, 6, 15, 15, 8

2. Construct a grouped frequency distribution for the following data in which the interval width (i) = 5 and the lower limits of the intervals are multiples of 5.
DATA: 50, 44, 36, 12, 23, 14, 8, 10, 5, 45, 36, 9, 9, 9

3. From the grouped frequency distribution in question 2, construct a frequency polygon. Use the midpoints on the abscissa and absolute frequency on the ordinate.

4. From the grouped frequency distribution in question 2, construct a histogram. Use the midpoints and relative frequency.

5. From the grouped frequency distribution in question 2, construct an ogive using the real limits and cumulative frequency.

6. State the exact limits for each of the following scores or intervals.

	Score or Interval	Exact Limits
a)	2.0 sec	__________
b)	2.6 sec	__________
c)	2–4 sec	__________
d)	25–29 kg	__________
e)	24.5–29.5 kg	__________
f)	24.50–29.50 kg	__________

7. The restaurant owner keeps track of the number of times that each of five dinner entrees is ordered over two 5-day periods, one in mid-winter and one in mid-summer. A total of 59 customers ordered one of the five dishes during this period in the winter and 47 in the summer. Which type of graph should be used to present these data? What measure of frequency should go on the ordinate?

8. Find the midpoint for the following intervals. Construct a frequency polygon using relative frequency.

X	f	Midpoint
23–25	1	
20–22	2	
17–19	2	
14–16	4	
11–13	5	
8–10	3	
5–7	1	
2–4	2	

9. Construct a scattergraph of the following data.

S#	X	Y
1	78	65
2	66	43
3	54	38
4	89	79
5	46	67
6	70	67
7	55	69
8	95	80
9	64	72
10	86	62

10. A psychiatrist has classified his 35 clients as either psychotic or neurotic. For each patient, he obtained a record of the number of pathological episodes that occurred in the past week. He wishes to present his data graphically so that he can easily compare the groups. What is type of graph should he use? Which measure of frequency should be used?

CHAPTER 3

Describing the Central Tendency of Distributions

Measures of central tendency describe the "average" of a distribution of scores. There are several ways of measuring "average." We will consider the three most commonly used measures; **mode, median** and **mean**.

Three Measures of Central Tendency

Mode

The mode (**Mo**) is the most frequently occurring value in a distribution. The mode is the score with a higher frequency than any other score in the distribution. It is the "typical" value in a distribution.

Median

The median (**Mdn**) is the score point at or below which, exactly 50% of the scores lie. The median divides the distribution of scores in half.

Mean

The mean is the arithmetic average obtained by summing all the scores and dividing the sum by the total number of scores in the distribution. The mean is the measure most of us think of when we talk about "average." Statisticians use the symbol μ "mew" to refer to the mean.

Calculating Measures of Central Tendency

Mode

The mode is easy to determine.

For Ungrouped Data

When data are ungrouped or in a simple frequency distribution, the mode is simply the value with the greatest frequency. A distribution may have more than one mode, and some distributions have no mode; differential frequencies must occur before a mode can be determined. The mode must have a higher frequency than other values. The following examples show numerical and graphic representations of various distributions.

	DATA	Mo
Unimodal (one mode)	1, 2, 3, 4, 5, 5, 6	5
Bimodal (two modes)	1, 2, 2, 4, 5, 5, 6	2, 5
No mode	1, 2, 3, 4, 5, 6 3, 3, 4, 4, 5, 5	all values occur an equal number of times.

Here are some examples of smoothed out frequency polygons.

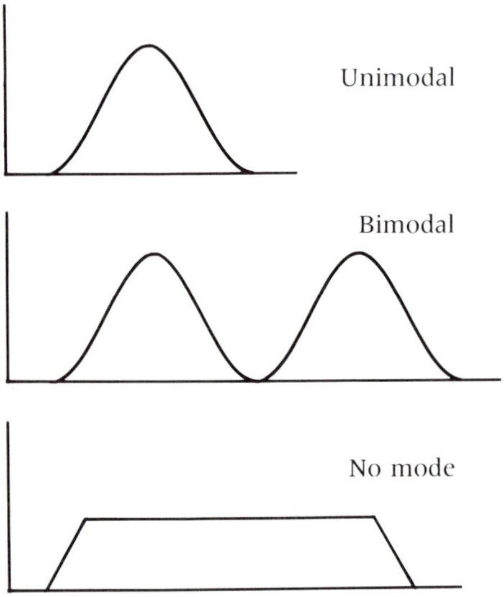

For Grouped Data

When the data have been placed in a grouped frequency distribution, the mode is the mid-point of the interval containing the highest frequency. The mode of grouped data is often called the "crude" mode, since the mid-point of the interval may or may not actually be the score value with the highest frequency. The mid-point, however, is used to represent the interval. Here is an example.

X	f	
50–54	3	
45–49	2	
40–44	7	Mo = 42
35–39	4	
30–34	3	
25–29	2	

The interval between 40 and 44 contains the highest frequency and so the mid-point of that interval (42) is the mode. As with ungrouped data, there must be differential frequency before a mode can be determined.

Median

Calculation of the median is complicated by certain conditions.

For Ungrouped Data

Median calculation for ungrouped data is a simple matter if there are no repeated values in the middle. If values are repeated, however, certain assumptions must be made.

Data with No Repeated Middle Values When the scores are arranged from highest to lowest, the median is the middle value, provided there are no other scores with this value. If there is an even number of scores, the median is the value halfway between the two middle values, providing again that these are not repeated. For example:

	Mdn
DATA: 1, 2, 3, 4, 5, 6, 7	4
1, 2, 3, 4, 5, 5, 5	4
1, 2, 3, 4, 5, 6, 7, 8	4.5

Linear Interpolation when Middle Values are Repeated When repeated middle values occur in a distribution, we must use **linear interpolation** to determine the median. Linear interpolation assumes that repeated values are *evenly distributed* between the exact or real limits of the score. For example, consider a score value of 3 that occurs twice. The exact limits of that score are 2.5 and 3.5. Our assumption requires that we distribute the two scores evenly between these limits. We do this by dividing the interval of one unit into two halves and placing one score in each half. We could draw this the following way:

.______X______.______X______. The X's represent the two scores.
2.5 3.5

Let's use this method to find the median when we have repeated middle values. Remember, the median is the score value at or below which *exactly* half the scores fall.

DATA: 1, 2, 2

Score	Exact Limits
1	0.5–1.5
2	1.5–2.5

Step 1. Draw a line representing the exact limits of the scores.

.__________.__________.
0.5 1.5 2.5

Step 2. Divide each interval into as many equal distances as you have scores.

.__________._____._____.
0.5 1.5 2.5

Between 0.5 and 1.5 we have only one score so we do not divide the interval. Between 1.5 and 2.5 we have two scores and so we divide the interval into two equal parts.

Step 3. Place the scores in the middle of each divided portion.

.____X____.__X__.__X__.
0.5 1.5 2.5

Step 4. Count exactly half the scores and locate the median.

1 1/2 below 1 1/2 above One half of 3 is 1 1/2

.____X____.__X__.__X__.
0.5 1.5 2.5
Mdn

Step 5. Add the lower exact limit of the interval containing the median to the exact portion of that interval needed to reach the median.

1/2 1/2
.____X____.__X__.__X__.
0.5 1.5 2.5

$$\text{Mdn} = 1.5 + 1/2(1/2) = 1.75$$

We need exactly one quarter of the interval to reach the median. Here are some more examples.

DATA: 1, 2, 2, 2, 3, 4

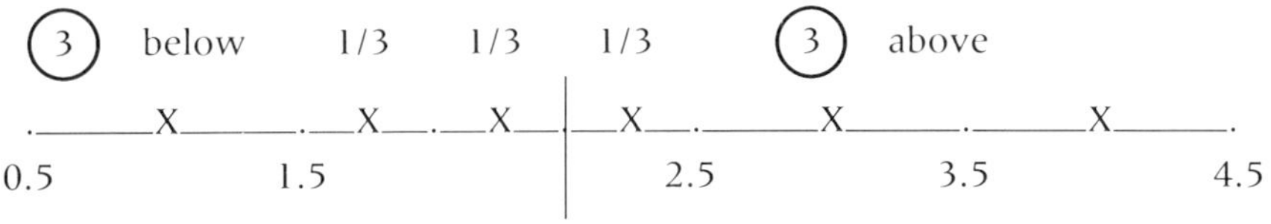

Mdn = 1.5 + 1/3 + 1/3 = 2.17

It is not necessary to draw the line representing all the numbers but only that portion where the median lies.

DATA: 5, 7, 8, 8, 8, 8, 8

We need 1/2 of 7 scores above and 1/2 below the median. This value must lie in the interval from 7.5 to 8.5.

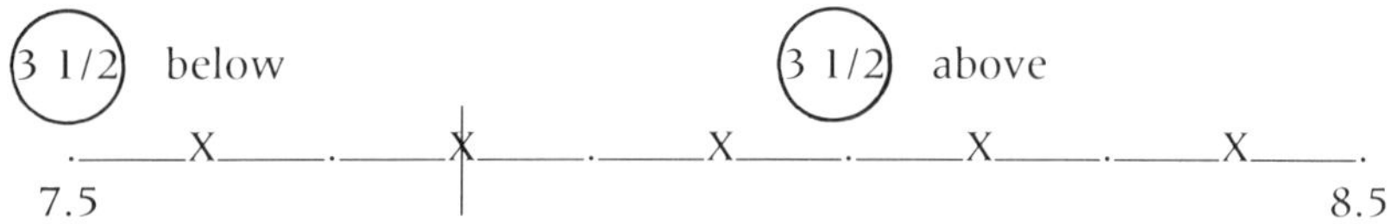

Mdn = 7.5 + 1/5 + 1/2(1/5) = 7.8

Two values are below 7.5. We need $1\frac{1}{2}$ more to reach the median.

For Grouped Data

Linear interpolation is also required to determine the median of grouped data. The procedure is basically the same. We must keep in mind, however, that the interval between exact limits is no longer one unit long. Here is an example.

Interval	f	cf
60–64	1	28
55–59	0	27
50–54	3	27
45–59	5	24
40–44	3	19
35–39	6	16
30–34	5	10
25–29	1	5
20–24	4	4

We have a total of 28 scores in this grouped frequency distribution. The median is the value at or below which exactly 14 scores fall. It lies somewhere in the interval between 35 and 39. We will draw our line to represent the exact limits of that interval.

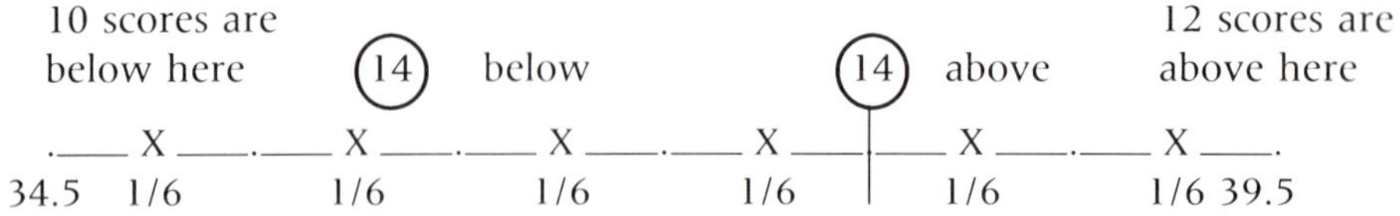

$$\text{Mdn} = 34.5 + 4/6(5) = 37.83$$

We need to add $^4/_6$ of the interval to reach the median. The interval is 5 units long in this case and so we need to add $^4/_6$ of 5 units to the exact lower limit of the interval.

Linear interpolation is always required when data are grouped or when ungrouped data have repeated values in the middle. A formula is available which performs the linear interpolation for you.

$$\text{Mdn} = L + \frac{\left[N\left(\frac{50}{100}\right) - f_b\right]i}{f_w} \qquad \text{Median}$$

L = lower exact limit of the interval containing the median.

f_b = total frequency below the lower limit of the interval containing the median

f_w = frequency within the interval containing the median. For ungrouped data this is the number of repeated middle values.

i = interval width. For ungrouped data the width is always 1.

Let's use the formula to verify the previous example.

$$\text{Mdn} = 34.5 + \frac{\left[28\left(\frac{50}{100}\right) - 10\right]5}{6}$$

$$= 34.5 + 4/6(5) = 37.83$$

To use the formula you must first find the interval containing the median. The portion of the formula to the right of the plus sign determines the proportion of the interval needed to reach the median. This formula can be used for calculating the median of grouped frequency distributions or for distributions which are ungrouped but have repeated values in the middle.

Mean

For Raw Data

When data are not in a frequency distribution, the mean is calculated by summing all the scores and dividing by the total number of them. The formula for the mean is

$$\mu = \frac{\Sigma X}{N} \quad \text{Mean for Raw Data}$$

X is a general symbol standing for all the scores in the distribution. The ΣX tells us to sum up all the scores in the X distribution. N is the total number of scores. Ready for an example?

DATA: 5, 4, 3, 4, 3, 0, 2
$\Sigma X = 21 \quad N = 7 \qquad \mu = 21/7 = 3$

I don't think we need to belabor this point, do you?

For Data in a Simple Frequency Distribution

When data are in a frequency distribution we must multiply each score by its frequency before we sum and then divide by the number of scores. The formula is

$$\mu = \frac{\Sigma fX}{N} \quad \text{Mean for Data in a Frequency Distribution}$$

Here is an example.

X	f	fX
5	1	5
4	2	8
3	2	6
2	1	2
1	0	0
0	1	0

$\Sigma fX = 21$
$N = \Sigma f = 7$
$\mu = 21/7 = 3$

For Data in a Grouped Frequency Distribution

When we determined the mode of a grouped frequency distribution, we used the mid-point of the interval with the highest frequency as our mode. Similarly, the mean is determined by using the midpoint of each interval to represent that interval.

Interval	Midpt(X)	f	fX
25–29	27	2	54
20–24	22	2	44
15–19	17	4	68
10–14	12	1	12
5–9	7	0	0
0–4	2	1	2

$$\Sigma fX = 180$$
$$N = \Sigma f = 10$$
$$\mu = 180/10 = 18$$

The mean of a grouped frequency distribution will not exactly equal the mean of the raw data because mid-points are used to represent the intervals, not the actual scores. The larger the interval width, the greater the amount of error due to grouping.

Interpreting Measures of Central Tendency

Mode

The mode is interpreted as the "typical" value in a set of scores. If we are interested in the value which occurs more often than any other, then the mode is appropriate. A dress manufacturer, for example, who wants to make dresses of only one size, would sell the most by using the modal dress size. This way he could fit the greatest number of people. The mode is a quick way of describing central tendency. The mode is the only measure of central tendency suitable for nominal data. Median or mean hair color makes no sense, but modal hair color does.

Median

The median is the middle point in a distribution. It divides the number of scores in a distribution exactly in half. If you want to divide a large number of scores into two equal-sized groups, the median would be a good place to do this.

Mean

Deviation score: The difference between a score and the mean of its distribution.

The mean can be considered the *balance point* of a distribution of scores. The scores below the mean "balance" the scores above the mean. A way to conceptualize the mean is to consider what happens when two children play on a teeter-totter. If one playmate is heavier than the other, he will have to sit closer to the middle of the teeter-totter to balance. If one playmate is lighter, then he must sit further toward the end to balance the board.

The mean acts like the fulcrum of a teeter-totter. The scores represent the weight of the children sitting on the teeter-totter. The distance of the scores above or below the mean represents the distance from the fulcrum that each playmate sits to make the board balance.

The distance between a particular score and the mean of the distribution is its **deviation score**. For example, if a score of 80 comes from a distribution with a mean of 65, then the deviation score for that raw score is 15. The raw score is 15 units away from the mean.

The formal way of expressing the mean as a balance point is $\Sigma(X - \mu) = 0$.

This equation says that if all the scores in a distribution are expressed in terms of deviations from the mean, then the sum of all the deviations is zero. For deviations taken about any mean, the sum of the negative deviations is numerically equal to the sum of the positive deviations.

Arithmetic and algebraic proofs of this appear below.

Arithmetic Proof:

DATA: 2, 3, 4, 5, 6

$$\mu = 20/5 = 4$$

$$\begin{aligned}\Sigma(X - \mu) &= (2 - 4) + (3 - 4) + (4 - 4) + (5 - 4) + (6 - 4)\\ &= -3 + 3\\ &= 0\end{aligned}$$

Algebraic Proof:

$$\begin{aligned}\Sigma(X - \mu) &= \Sigma X - \Sigma\mu\\ &= \Sigma X - N\mu^{1}\\ &= \Sigma X - N(\Sigma X/N)\\ &= \Sigma X - \Sigma X\\ &= 0\end{aligned}$$

[1] The sum of a constant is the constant multiplied by the number of times it appears in the distribution, i.e., N.

Let's go back to our teeter-totter analogy. The line below represents the teeter-totter. The fulcrum is the mean. The raw scores and their deviations are indicated.

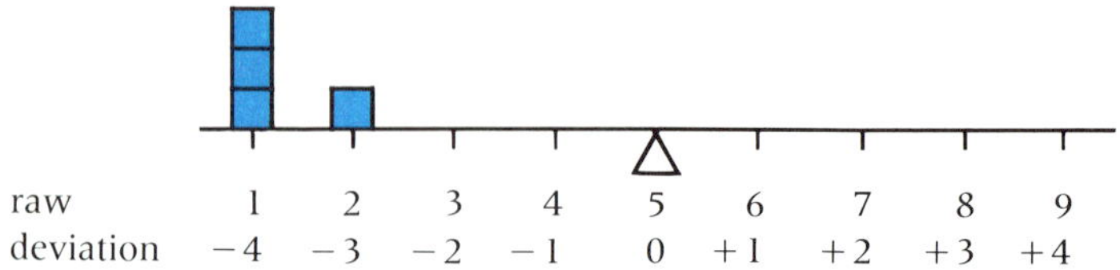

In this example, there are three raw scores with a value of 1. They are 4 deviations below the mean; their total deviation value is -12 units, -4 units for each score. Another raw score is a 2 which is 3 units below the mean; its deviation score is -3. The total of the negative deviations is $(-12) + (-3) = -15$.

To balance we must have equal "weight" above the mean, so we need a total of $+15$ deviations. There are several ways this could occur. For example, we could place 15 scores at the raw score value of 6. This would produce a total of $15\,(+1) = +15$ deviations and the board would balance. Could we balance the board by adding only one score?

Yes: It would have to provide $+15$ deviations so it would have to be 15 units above the mean. Since the mean is 5, our single score would have a value of 20 to balance the distribution.

The sum of the deviations is determined by their size and frequency. A single score far away from the mean may be balanced by many scores close to the mean, on the other side.

Shape and Measures of Central Tendency

Symmetrical Distributions

Recall that in a symmetrical distribution the right half, the side above the mean, is a mirror image of the left half, the side below the mean. For any symmetrical distribution, the mean and median are equal. If the distribution is also unimodal, the mean, median and mode are equal. Here are some examples.

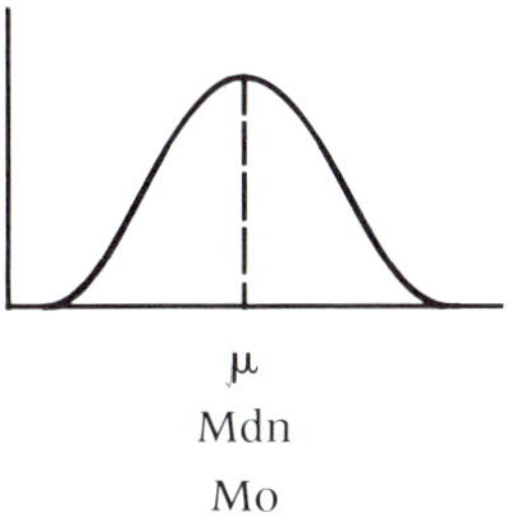

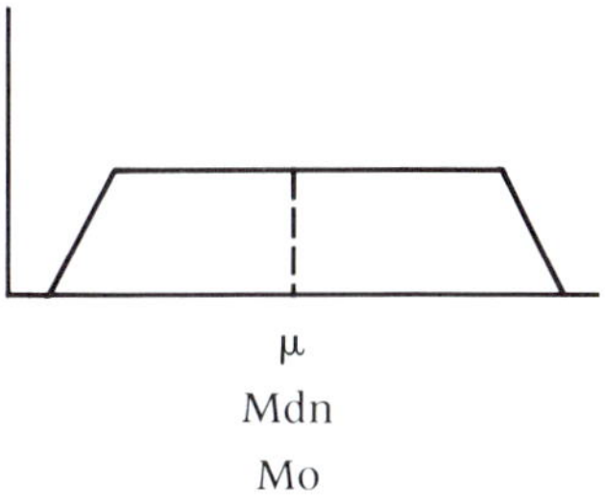

Skewed Distributions

Recall that skewed distributions are labeled on the basis of the direction of the tail. When the tail points to the negative or left side of the distribution, it is called a negative skew. A positively skewed distribution has the tail pointing to the right or positive side. When a distribution is skewed, the mean, mode and median are not equal. The mean always lies closest to the tail. The mode is always at the highest point and the median is always between the mean and mode. Here are some examples.

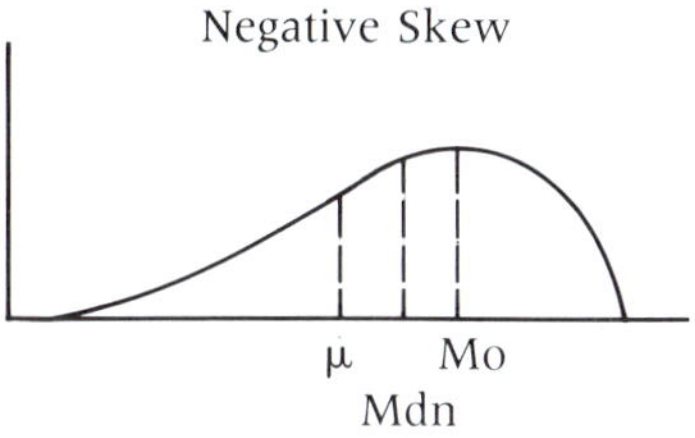

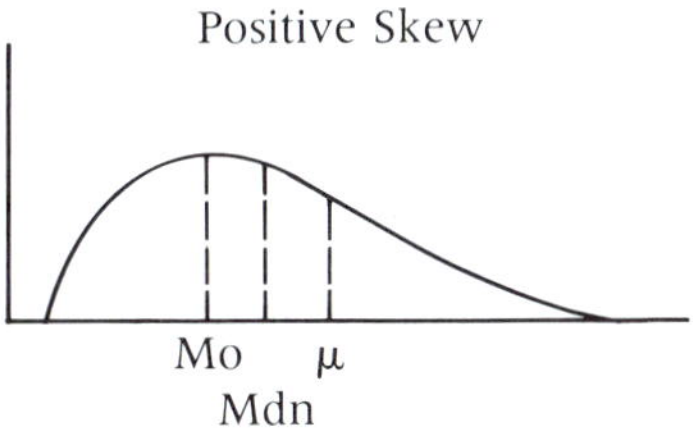

Comparing Measures of Central Tendency

Choosing which of the three measures of central tendency to use as your "average" depends on your reasons for reporting the measure, the kind of data you have, and what additional measures you will be taking.

Sensitivity to Score Position in a Distribution

The mean is most sensitive to the exact value of each score in a distribution because all the scores are used to compute the mean. If your objective is to reflect all the scores, then the mean is the most appropriate "average" to use. If, however, you do not want your measure of "average" to be "pushed"

Open-ended distribution: One where the exact upper or lower limit is unknown.

toward a few very extreme scores, then the median may be a better choice. The median is not sensitive to the exact value of each score but only to the number of scores above and below it. It doesn't matter how far above or how far below those scores are. In the example below, the median is the same although the distributions are clearly different.

Distribution A: 1, 10, 100, 1000, 10 000

Distribution B: 98, 99, 100, 101, 102

The median of both distributions is 100, but the mean of distribution A is 2222.2 and the mean of distribution B is 100. The median is not sensitive to score value but only to score frequency. The mean, on the other hand, is sensitive to score values because its computation uses each score value. This requires, of course, that we know the value of each score!

Sometimes we find ourselves in a situation where we do not know each score. In these cases the median may be the only "average" we can determine. For example, consider an experiment where 100 rats are timed running through a maze. After the experiment is complete, we find that 12 rats got lost in the maze and didn't finish. If we wish to include those rats in our measure of "average running time," we cannot use the mean. We could, however, compute the median. We would count those12 rats as taking more time than even the slowest rat. This would produce an **open ended distribution**; there are no values for the extreme end. The median can be calculated for most open-ended distributions since it relies on the number of cases above and below it, not on the distance.

The mode is not sensitive to exact score values either. It is sensitive only to the score with the highest frequency. It is often used as a very quick and somewhat crude measure of central tendency. The mode is the only measure of central tendency appropriate for nominal variables.

Resistance to Sampling Fluctuation

Inferential statistics use samples drawn from a population to make inferences about the properties of that population. We assume that the characteristics of the sample adequately reflect the characteristics of the population from which the sample was drawn. If samples were repeatedly drawn from a population, and the three measures of central tendency computed for each sample, the values of the mean would be more similar than the medians or modes. The mean fluctuates the least from sample to sample. The greatest variation would occur in the modes. The importance of this will become evident in later chapters.

The Mean of Combined Subgroups

Combined mean: The grand mean of all the scores in all groups when two or more groups are combined.

Frequently it is necessary to find the overall mean of two or more distributions of scores. For example, by the end of the year my statistics students have written six exams. I always determine the mean on each exam as I record the grades. What if I would like to know the overall mean of the class on the first two exams or all six? If the same number of students wrote each exam, I would simply compute the mean of the means. Unfortunately, even in *my* statistics class, some students drop out during the year and so the number taking each exam often differs. Let's use the data from two exams to illustrate how to compute the **combined mean** of two subgroups.

If we have the sum of the raw scores for each exam, we would use the following formula to compute the combined mean (μ_c).

$$\mu_c = \frac{\Sigma X + \Sigma Y}{N_x + N_y}$$

Mean of combined subgroups when the sums are known.

The X distribution will be the scores of the 20 students who wrote the first statistics exam. The Y distribution will be the scores of the 18 who remained to write the second exam.

$$\Sigma X = 1302$$
$$\Sigma Y = 1070$$
$$\mu_c = \frac{1302 + 1070}{20 + 18} = 62.42$$

Let's say that we don't happen to have the sums of the raw scores on the exams but we do have the means for each. In this case we would use the following formula.

$\mu_c = \dfrac{N_x\mu_x + N_y\mu_y}{N_x + N_y}$	Mean of combined subgroups when the means are known.

The mean on the first exam was 65.10 and the mean on the second exam was 59.44.

$$\mu_x = 65.10$$
$$\mu_y = 59.44$$

$$\mu_c = \frac{20(65.10) + 18(59.44)}{20 + 18}$$
$$= 62.42$$

The mean for each exam was computed using a different number of students. There were 20 students who wrote the first exam, but only 18 wrote the second. We cannot simply average the two means because they were computed on different group sizes. We must first weight each mean by the number of cases that contributed to its calculation, sum these weighted scores, and then divide by the total number of scores. As you can see, this weighting procedure produces the subgroup sums:

$\mu = \Sigma X/N$ and by rearranging things $\Sigma X = N\mu$

When the subgroups are equal in size, we don't have to weight the means but simply average the means by adding them and dividing by the number of subgroups. If all 20 students had written the second exam and its mean was 59.44, then we could find the combined mean in the following way.

$$\mu_c = \frac{\mu_x + \mu_y}{2} = \frac{65.10 + 59.44}{2} = 62.27$$

Although this value is close to the value we got when we used the weighted mean formula, it is not exactly the same. The amount of error produced by averaging the means of different-sized groups depends on how different the group sizes are. Larger differences produce larger error.

Measures of central tendency are important techniques for describing the average of distributions. In the next chapter, we will study another important characteristic of distributions: spread or variability.

SUMMARY OF TERMS AND FORMULAS

The **mode** is the most frequently occurring value in a distribution.

The **median** is the score value at or below which 50% of the values fall.

The **mean** is the arithmetic average of all the scores.

Measure	Formula	
Median	$\text{Mdn} = L + \dfrac{\left[N\left(\frac{50}{100}\right) - f_b\right]i}{f_w}$	
Mean for Raw Data	$\mu = \dfrac{\Sigma X}{N}$	
Mean for Data in a Frequency Distribution	$\mu = \dfrac{\Sigma fX}{N}$	
Mean for Combined Subgroups	$\mu_c = \dfrac{\Sigma X + \Sigma Y}{N_x + N_y}$	when sums are known
	$\mu_c = \dfrac{N_x\mu_x + N_y\mu_y}{N_x + N_y}$	when means are known
	$\mu_c = \dfrac{\mu_x + \mu_y}{2}$	when $N_x = N_y$

EXERCISES

1. Calculate the mean, mode and median for the following sets of raw data. Use linear interpolation for calculating the medians.

DATA SET A: 10, 5, 8, 7, 3, 5, 9, 8, 3, 8, 11
DATA SET B: 5, 3, 8, 2, 1, 0, 2

2. For the two distributions in question 1, determine the combined mean.

3. For the two distributions in question 1, draw a diagram that indicates the skew for each set of data.

4. Calculate the mean, mode and median for the following two sets of data. Use the formula for the median calculations. What direction is the skew for each data set?

DATA SET A:

X	f
10	3
9	1
8	2
7	2
6	5
5	6

DATA SET B-Grouped F.D:

Mid-points	f
152	2
147	1
142	2
137	5
132	7
127	9
122	8
117	13
112	11
107	14
102	12
97	4
92	5
87	5
82	2

5. For the following data, calculate the mean, mode and median. What direction is the skew?

X	f
16–18	5
13–15	3
10–12	7
7–9	6
4–6	2
1–3	4

6. A distribution has a mean of 10. The sum of the positive deviations is +60. If the sum of the negative deviations of all but one score is −45, what is the value of the last score?

7. Income is a positively skewed distribution.

a) Why do you think this is so?

b) If Statistics Canada wants to impress the world with our standard of living, which measure of central tendency should they use to report average income of Canadians?

CHAPTER 4

Describing the Variability of Distributions

Chapter 3 dealt with measures of central tendency—methods for indicating the "average" of a distribution. Measures of **variability** provide methods for describing the variation in scores or the amount of scatter around the centre of a distribution. These measures indicate whether scores are clustered closely around the middle of the distribution or whether they are scattered far from the middle. To adequately describe a distribution of scores, a measure of variability as well as a measure of central tendency is required. There are several ways of measuring variations in scores. We will consider three of these: **range, variance,** and **standard deviation**.

Three Measures of Variability

Range

The range of a distribution is the number of values over which the distribution spans. The range gives us a quick measure of variability.

Variance

The variance is the average of the squared deviations of scores from the mean of the distribution. The symbol for the variance is σ^2 "sigma squared."

Standard Deviation

The standard deviation is used most often to describe variability. It is the square root of the variance and its symbol is σ "sigma."

Calculating Measures of Variability

Range

The range is by far the easiest measure to determine. It is the difference between the highest and lowest score, plus one. In a distribution of test scores, if someone got 1% and someone else got 100%, then the range would be $100 - 1 + 1 = 100$. In other words, the scores would range over a total of 100 possible values. We determined the range in Chapter 2, when we constructed a grouped frequency distribution. This measure can be thought of as the *width* of the entire distribution.

Variance and Standard Deviation

Calculating these two measures is not difficult, however, it is somewhat tedious. Before we begin, let's examine some symbols and abbreviations.

The previous chapter discussed the mean as a balance point because the sum of the negative deviations equals the sum of the positive deviations. We will need the concept of the deviation score again, this time to calculate variance and standard deviation. We will use a new symbol to stand for deviation score. I call this "squiggley x" (x).

Defining Formulas

The deviation of a score from its mean is determined by subtracting the mean from the score. This operation can be expressed as $X - \mu = x$.

The variance is the average of the squared deviations of the scores from the mean of the distribution. The formula which defines the variance is:

$$\sigma^2 = \frac{\Sigma x^2}{N} = \frac{\Sigma(X - \mu)^2}{N}$$

Population Variance: Defining Formula

any x value can substitute in

To use the variance formula we follow:

Step 1. Subtract the mean from each raw score.

Step 2. Square each difference.

Step 3. Sum all the squared differences.

Step 4. Divide this sum by N (the total number of cases).

The standard deviation is simply the square root of the variance and so to determine the standard deviation, we add:

Step 5. Take the square root of the number obtained in step 4.

The formula which defines the standard deviation is

$$\sigma = \sqrt{\frac{\Sigma x^2}{N}}$$

Population Standard Deviation: Defining Formula

Let's use an example to show the steps for calculating the variance and the standard deviation of a distribution of scores.

X	x	x^2
1	−9	81
4	−6	36
7	−3	9
13	+3	9
16	+6	36
19	+9	81
$\mu = 10$		$\Sigma x^2 = 252$

$$\sigma^2 = \frac{\Sigma x^2}{N} = \frac{252}{6} = 42$$

$$\sigma = \sqrt{42} = 6.48$$

The middle column of numbers contains the deviation scores for each raw score. These deviations are then squared in the right hand column. The sum of the squared deviations is divided by N for the variance. The standard deviation is the square root of the variance.

Computational Formulas

The defining formulas for these two measures of variability are helpful because they show us exactly what we are doing when we compute a variance or standard deviation. However, they are not used often in calculation. The computational formulas for variance and standard deviation have been derived from the defining ones and are easier to use.

For Raw Data When data have not been organized into a frequency distribution, but appear as single raw scores, the formulas for the variance and standard deviation are

$$\sigma^2 = \frac{\Sigma X^2 - (\Sigma X)^2/N}{N}$$ Variance for Raw Data

$$\sigma = \sqrt{\frac{\Sigma X^2 - (\Sigma X)^2/N}{N}}$$ Standard Deviation for Raw Data

ΣX^2 is the sum of the squared raw scores. To calculate ΣX^2:

1. Square all the scores.
2. Sum the squared scores.

$(\Sigma X)^2$ is the squared sum of the raw scores. To calculate $(\Sigma X)^2$:

1. Sum all the scores.
2. Square the sum.

Ready for an example? We will use the data we used earlier when we computed variance and standard deviation with the defining formulas.

X	X^2
1	1
4	16
7	49
13	169
16	256
19	361
$\Sigma X = 60$	$\Sigma X^2 = 852$

$$\sigma^2 = \frac{\Sigma X^2 - (\Sigma X)^2/N}{N}$$

$$= \frac{852 - 60^2/6}{6} = 42$$

$$\sigma = \sqrt{42} = 6.48$$

For Data in a Simple Frequency Distribution When data have been organized into a frequency distribution, the computational formulas for the variance and standard deviation are

$$\sigma^2 = \frac{\Sigma fX^2 - (\Sigma fX)^2/N}{N}$$ Variance for Data in a Frequency Distribution

$$\sigma = \sqrt{\frac{\Sigma fX^2 - (\Sigma fX)^2/N}{N}}$$ Standard Deviation for Data in a Frequency Distribution

When we calculated the mean for data in a frequency distribution, we multiplied each score by its frequency. We must do the same here.

To calculate ΣfX^2:

1. Square each X value.
2. Multiply each square by its frequency.
3. Sum all these products.

To calculate $(\Sigma fX)^2$:

1. Multiply each X value by its frequency.
2. Sum these products.
3. Square the sum.

For Data in a Grouped Frequency Distribution When calculating the variance and standard deviation of data in a grouped frequency distribution, we represent each interval by its midpoint just as we did earlier when computing the mean. This procedure is somewhat crude and does not always produce the same result as the raw data. Since the extensive use of calculators this method may be obsolete. I include it because there may be occasions when you do not have access to the original data. Here is an example.

Interval	M.P.	f	fX	X^2	fX^2
155–159	157	1	157	24 649	24 649
150–154	152	2	304	23 104	46 208
145–149	147	1	147	21 609	21 609
140–144	142	7	994	20 164	141 148
135–139	137	4	548	18 769	75 076
130–134	132	5	660	17 424	87 120
125–129	127	3	381	16 129	48 387
120–124	122	2	244	14 884	29 768
115–119	117	1	117	13 689	13 689
		26	3552		487 654

$$\sigma^2 = \frac{\Sigma fX^2 - (\Sigma fX)^2/N}{N}$$

$$= \frac{487\,654 - (3552)^2/26}{26}$$

$$= 92.16$$

$$\sigma = \sqrt{92.16} = 9.60$$

Interpreting Measures of Variability

Range

The range indicates the total number of possible values over which the raw scores span. Since the range uses only the highest and lowest scores in its calculation, it is a relatively crude measure of variability. When the median has been used as the measure of central tendency of a distribution, the range may be the most appropriate measure of variability.

Variance

The variance is a number reflecting the "average" squared distance or deviation of scores from the mean of the distribution. Since the deviations are squared, the variance can be hard to interpret because it is not in the same arithmetic units as the raw scores.

Standard Deviation

Like the variance, the standard deviation indicates the "average" distance or deviation of the scores from the mean. Because the standard deviation is the square root, it is in the same units as the raw scores and is easier to interpret. A standard deviation of 6, for example, means that, on the average, the scores are located 6 points from the mean.

The standard deviation is a quick way of locating the bulk of the scores in a distribution. In any distribution, regardless of the shape, at least 75% of the scores are within two standard deviations of the mean and at least 89% are within three standard deviations of the mean. A distribution, then, with a standard deviation of 6, has at least 75% of its scores between 12 points below and 12 points above the mean, and 89% between 18 points below and above the mean.

For a fairly symmetrical distribution, more than 90% of the observations fall within two standard deviations of the mean and almost all the observations are within three standard deviations of the mean.

For a normal distribution, which we will study further in later chapters, the following is true:

$$\mu \pm 3\,\sigma = 99.74\%$$

$$\mu \pm 2\,\sigma = 95.44\% \qquad \text{In the Normal Distribution}$$

$$\mu \pm 1\,\sigma = 68.26\%$$

This means that in a normal distribution, 99.74% of the scores fall between three standard deviations above and below the mean; 95.44% of the scores are within two standard deviations; 68.26% of the scores are within one standard deviation of the mean. These facts will provide a basis for several statistical techniques that we will study in the chapters on inference.

Whenever we describe a distribution of raw scores, we should include a measure of variability along with a measure of central tendency. It is not sufficient to state the "average" only. For example, two distributions may have the same "average" but are very different in terms of spread. The following illustration shows two distributions of scores whose means are identical. The A distribution, however, has scores clustered around the centre, whereas the B distribution has scores spread far from the centre. To adequately describe these two distributions, we must include some measure of spread as well as a measure of central tendency.

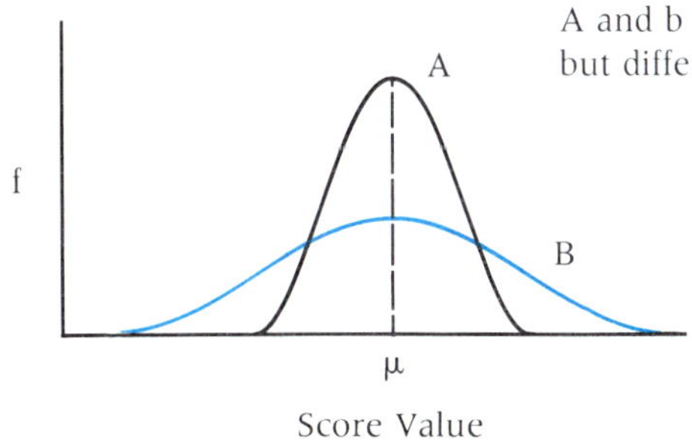

A and b have the same mean
but different spreads

When the median has been used as the measure of central tendency, it is customary to report the range as the variability measure. When the mean is used, the standard deviation usually accompanies it.

The Variance and Standard Deviation of Combined Subgroups

Pooled variance and standard deviation: The variance and standard deviation of all the scores in all groups around the combined mean when two or more groups are combined.

When we studied central tendency, we discovered that finding the combined mean of two or more groups is often not as simple as taking the average of the means of each group. Finding the overall variance and standard deviation of several subgroups is no easy task either. In fact, it can be downright complicated.

Variance and standard deviation are measures of spread around the mean of a distribution of scores. To find the variance or standard deviation of combined subgroups, we must determine the average spread of *all the scores around the combined mean.* We cannot simply average the subgroups' variances or standard deviations because each refers to the spread of group scores around the mean of that group. Rather, we must determine the spread of *all* the scores from *all* the distributions around the mean of *all* the scores to find the **pooled variance** or standard deviation. The formula for determining the pooled variance of three groups is presented below. This formula may be modified for any number of subgroups. The square root of the pooled variance gives the pooled standard deviation.

Combined or pooled variance:

$$\sigma^2_c = \frac{N_w\sigma_w^2 + N_x\sigma_x^2 + N_y\sigma_y^2 + N_w(\mu_w - \mu_c)^2 + N_x(\mu_x - \mu_c)^2 + N_y(\mu_y - \mu_c)^2}{N_w + N_x + N_y}$$

Each subgroup variance must be weighted by the number of scores in the group and the squared deviation of each subgroup mean from the combined mean must also be weighted by the group size. To illustrate this, let us determine the pooled variance for the following data:

$$\mu_w = 8 \qquad \sigma_w^2 = 2 \qquad N_w = 10$$
$$\mu_x = 6 \qquad \sigma_x^2 = 1 \qquad N_x = 12$$
$$\mu_y = 10 \qquad \sigma_y^2 = 3 \qquad N_y = 15$$

The first step is to determine the combined mean discussed in Chapter 3. With three groups the formula for the combined mean is

$$\mu_c = \frac{N_w\mu_w + N_x\mu_x + N_y\mu_y}{N_w + N_x + N_y} = \frac{10(8) + 12(6) + 15(10)}{10 + 12 + 15}$$

$$= 302/37 = 8.16.$$

Now we can go ahead and compute the combined variance.

$$\sigma^2_c = \frac{N_w\sigma_w^2 + N_x\sigma_x^2 + N_y\sigma_y^2 + N_w(\mu_w - \mu_c)^2 + N_x(\mu_x - \mu_c)^2 + N_y(\mu_y - \mu_c)^2}{N_w + N_x + N_y}$$

$$= \frac{10(2) + 12(1) + 15(3) + 10(8 - 8.16)^2 + 12(6 - 8.16)^2 + 15(10 - 8.16)^2}{10 + 12 + 15}$$

$$= \frac{77 + 107.03}{37} = 4.97$$

Notice that the combined variance is *not an average* of the subgroup variances. Rather it is a measure of the spread of *all* the subgroup scores around the combined mean.

SUMMARY OF TERMS AND FORMULAS

Measures of variability describe the spread or scatter of the scores in a distribution.

The **range** measures the span of the distribution. The **variance** and **standard deviation** measure how far, on the average, the scores are from the mean of the distribution.

When the **mean** is used as the measure of central tendency, the **standard deviation** usually accompanies it as the measure of variability. When the **median** is used, it is usually accompanied by the **range**.

Measure	Computational	Defining
Range	$H - L + 1$	
Variance for Raw Data	$\frac{\Sigma X^2 - (\Sigma X)^2/N}{N}$	$\frac{\Sigma(X - \mu)^2}{N}$
Variance for Data in a Frequency Distribution	$\frac{\Sigma fX^2 - (\Sigma fX)^2/N}{N}$	
Standard Deviation for Raw Data	$\sqrt{\frac{\Sigma X^2 - (\Sigma X)^2/N}{N}}$	$\sqrt{\frac{\Sigma(X - \mu)^2}{N}}$
Standard Deviation for Data in a Frequency Distribution	$\sqrt{\frac{\Sigma fX^2 - (\Sigma fX)^2/N}{N}}$	

Pooled Variance for Three Combined Subgroups

$$\sigma^2_c = \frac{N_w\sigma_w^{\ 2} + N_x\sigma_x^{\ 2} + N_y\sigma_y^{\ 2} + N_w(\mu_w - \mu_c)^2 + N_x(\mu_x - \mu_c)^2 + N_y(\mu_y - \mu_c)^2}{N_w + N_x + N_y}$$

EXERCISES

1. For the following distributions, determine the deviation scores for each raw score.

a) 2, 4, 6, 7, 4 3, 2
b) 12, 23, 50, 35, 20

2. Use the defining formulas to calculate the variance and standard deviation of the following sets of data. What are the ranges?

DATA SET A: 10, 8, 4, 4, 8, 5, 2
DATA SET B: 6, 4, 8, 2, 1, 0, 0, 3

3. Use the computational formulas to calculate the variance and standard deviation for the following two sets of data. What are the ranges?

DATA SET A: 6, 4, 8, 2, 1, 0, 0, 3
DATA SET B:

X	f
10	3
9	1
8	2
7	2
6	5
5	6

4. Calculate the variance and standard deviation for the following grouped frequency distribution.

X	f
23–25	1
20–22	2
17–19	2
14–16	4
11–13	5
8–10	3
5– 7	1
2– 4	2

5. For the data of question 2, compute the pooled variance and standard deviation.

6. In which of the following distributions of test scores would you prefer to be with a score of 112? Explain your reasons.

Distribution A Mean = 98 Standard Deviation = 4
Distribution B Mean = 98 Standard Deviation = 16

CHAPTER 5

Describing the Position of Scores in Distributions

Frequently, we wish to describe the location of one score in a distribution of many scores. We will examine two techniques that are commonly used to do this.

Percentiles and **percentile ranks** use percentage to describe score position. After taking a test, you may wish to know how well you did compared to the rest of the class. You might find, for example, that your performance on a test was as good or better than 80% of the class. Your test score would have a percentile rank of 80.

Alternatively, you might ask "What score did I need to be in the top 10% of the class?" You might find that to be in the top 10% you would have had to score 78 points on the test. The score of 78, then, would be the 90th percentile.

These techniques rely on how many cases are above and below certain points in the distribution of scores. Recall that the median also does this by indicating the point at or below which half the cases fall; the median is the 50th percentile.

Another way of locating the position of a score is to compare that score with the mean of the distribution. **Z-scores** do this by indicating how far above or below the mean a particular raw score value lies. Z-scores rely on

> **Derived score:** Scores obtained by transforming raw scores.

distance from the mean; percentiles and percentile ranks rely on number of cases.

Percentile ranks and Z-scores are **derived scores** since they involve numbers which are derived from the raw scores in the distribution. They help us determine the relative standing of a particular score in a distribution of scores.

Percentiles and Percentile Ranks

Percentiles

A percentile is a *score point* at or below which a particular percentage of cases fall. $\mathbf{P_{PR}}$ is the abbreviation for percentile. The subscript indicates the % value at or below which an unknown score falls. A percentile can range over the possible values in the distribution. For example, P_{80} is the score value in a distribution at or below which 80% of the cases fall. P_{50} is the score value at or below which 50% of the cases fall. P_{50} is also the median of a distribution.

Percentile Ranks

The percentile rank ($\mathbf{PR_x}$) of a score is a *number* indicating the percentage of cases falling at or below a particular score value. It is a *percentage value* and can only range from 0 to 100. The subscript x is the score value and the PR indicates what percentage of the distribution is at or below that score value. For example, if the percentile rank of a score of 115 is 65 (PR_{115} = 65), then we know that 65% of the cases lie at or below the score of 115. Alternatively, we could say that 115 is the 65th percentile (P_{65} = 115).

Earlier, we found that we could divide a distribution exactly in half using the median or the 50th percentile. Similarly, we could divide a distribution into quarters using the 25th, 50th, 75th and 100th percentiles. This would produce four equal-sized groups. This is a common technique and these four percentiles are often called **quartiles**.

Calculating Percentiles and Percentile Ranks

Percentiles

The general formula for computing percentiles from a grouped frequency distribution is

$$P_{PR} = L + \frac{\left[N\left(\frac{PR}{100}\right) - f_b\right]i}{f_w} \qquad \text{Percentile}$$

where L is the lower exact limit of the score or interval containing the percentage of cases of interest; N is the total number of cases in the distribution; f_b is the number of cases below the score or interval containing the percentile; f_w is the number of cases within the interval containing the percentile; and i is the width of the interval.

This formula is almost the same as the one we used for calculating the median because the median is the 50th percentile. The procedure we will follow is the same as we used when computing the median. Ready for an example?

Below is a grouped frequency distribution.

Interval	f	cf	
155–159	1	25	
150–154	0	24	
145–149	2	24	P_{90}
140–144	2	22	
135–139	3	20	
130–134	4	17	
125–129	6	13	
120–124	4	7	
115–119	2	3	
110–114	1	1	

Let's determine P_{90}. When we were finding medians the first step was to determine where half the cases fell. To determine P_{90}, we must find where 90% of the cases fall.

Step 1. Determine what 90% of the total is. 90% of 25 is 22.5.

Step 2. Find the interval containing the 22.5th score. Looking at the cf column, we see that this score is in the interval 145–149.

Step 3. Determine the lower limit of this interval and substitute into the formula.

$$P_{90} = 144.5 + \frac{\left[25\left(\frac{90}{100}\right) - 22\right]5}{2}$$

$$= 144.5 + \frac{(22.5 - 22)5}{2} = 145.75$$

Therefore, 145.75 is the point in the distribution at or below which 90% of the scores fall.

Earlier we calculated the median by formula and by linear interpolation. Since the median is a percentile, we can use linear interpolation to find any other percentile. Using the above example, let's verify that the 90th percentile is 145.75.

22.5 below — 22.5 above

._______X_________._______X_______.

144.5 — 149.5

$$P_{90} = 144.5 + 1/4(5) = 144.5 + 1.25 = 145.75$$

We can see that there are two scores within the interval containing the 90th percentile. We need exactly one half of one of these scores so that there are 22.5 above and below. We need, then, one quarter of the entire interval which is 5 units long. This method can be used to determine any percentile for a grouped frequency distribution.

Percentile Ranks

The formula for determining the percentile rank of any score is

$$PR_x = \frac{\left[f_w\frac{(X - L)}{i} + f_b\right]100}{N} \qquad \text{Percentile Rank}$$

where X is the score value whose rank you wish to determine; L is the lower exact limit of the interval in which the score falls; f_w is the number of scores in that interval; f_b is the number of scores below that interval; i is the interval width; and N is the total number of scores in the distribution.

To illustrate this, let's use the same example we used earlier. Recall that the 90th percentile was 145.75. This means that the percentile rank of 145.75 must be 90. Let's use our percentile rank formula to prove that this is true.

$$PR_{145.75} = \frac{\left[\frac{2(145.75 - 144.5)}{5} + 22\right]100}{25}$$

$$= \frac{\left[\frac{2.50}{5} + 22\right]100}{25} = 22.5(4) = 90$$

Percentile ranks are derived scores that provide information about relative standing. They indicate the percentage of cases that are found at or below a particular score value. They do not, however, indicate the distance of the scores from the score in question. For example, a score with a percentile rank of 80 indicates that 80% of the cases were at or below that score but not how far below those cases were. A derived score, which indicates the actual distance a raw score is from the mean of a distribution, measured in units of standard deviation, is the Z-score.

The Z-Score

The Z-score (pronounced Zee in the U.S. and Zed in Canada) is a derived score that indicates how many standard deviations a raw score is from the mean and in what direction. This derived score is often more meaningful that the percentile rank of a score because it tells us how far above or below the mean the score value is. A raw score value above the mean will have a positive Z-score. A raw score below the mean will have a negative Z-score. A Z-score of − 1.00, for example, tells us that the raw score equivalent was exactly one standard deviation unit below the mean of the distribution.

Calculating Z-Scores

The general formula for determining the Z-score equivalent of any raw score is

$$Z = \frac{x}{\sigma} \qquad \text{Z for a Raw Score}$$

where $x = X - \mu$.

This formula simply divides the deviation score by the standard deviation of the distribution. In this way the Z value tells us whether the raw score was above or below the mean and exactly how far in standard deviation units.

Consider the distribution below. Its mean is 50 and its standard deviation is 7. Each raw score has been translated into its Z-score equivalent, using our formula.

Raw Scores	x	Z
68	18	2.57
54	4	0.57
51	1	0.14
50	0	0.00
50	0	0.00
48	−2	−0.29
48	−2	−0.29
46	−4	−0.57
45	−5	−0.71
40	−10	−1.43
$\bar{X} = 50$		0.00
$\sigma = 7$		

Notice that a raw score of 40 is nearly 1 1/2 standard deviation units below the mean. A score of 54 is more than 1/2 a standard deviation above the mean. Since Z indicates distance from the mean, a score which is equal to the mean, in this case 50, will always have a Z-score equivalent of zero, regardless of the standard deviation. Notice that the sum of the Z-scores is zero. This is true of any distribution of Z-scores. You will recall that the sum of the deviations about any mean is zero; since Z-scores are deviation scores divided by the standard deviation (a constant), their sum must also be zero.

Properties of a Z-Distribution

A **Z-distribution** is a distribution where all raw scores have been converted to their Z-score equivalents. All Z-distributions have certain unchanging properties.

The Mean

The mean of any Z-distribution is zero. This makes sense because the mean of the distribution does not deviate from itself. Formally, then:

$$\mu_Z = 0 \qquad \text{Mean of a Z-Distribution}$$

The Variance and Standard Deviation

The variance and standard deviation of any Z-distribution are always equal to 1. Formally:

$$\sigma_Z^2 = \sigma_Z = 1 \qquad \text{Variance and Standard Deviation of a Z-Distribution.}$$

The Shape

When a distribution of raw scores is converted to Z-scores, its shape *does not change.*

This is not true of all derived scores. The "dreaded" stanine system, used in many colleges and universities, changes the shape of the raw score distribution to a bell-shaped curve. The grading scheme used at my college changes the shape of the raw score distribution to a negatively skewed curve. Converting to Z, however, has no effect on the shape.

Using Z to Locate Scores in a Distribution

The Z-score is extremely useful for indicating the location or the relative standing of a score in a distribution of scores. Let's consider an example. Elmer obtained 80% on his psychology midterm exam. He was quite pleased with this grade. He was not so pleased, however, when he received 40% on his anthropology midterm. Although, at first glance, it is tempting to believe that 80% is the better grade, we are statistically sophisticated enough now not to jump to conclusions. First we must ask a couple of questions. We need to know the mean and standard deviation of each exam. Consider the following data:

Psychology Midterm	Anthropology Midterm
$\mu = 70\%$	$\mu = 35\%$
$\sigma = 20$	$\sigma = 2$
Elmer's score 80%	40%

On the anthropology test, most of the students were clustered around the mean and few got high marks. On the psychology test, the students scores were much more variable. Several students got high marks. It is difficult to evaluate Elmer's grades by glancing at the data. Certainly, he was above the mean in both cases, but how far? This question can be answered by converting his raw scores to Z scores.

$$\text{Z in psychology} = \frac{X - \mu}{\sigma} = \frac{80 - 70}{20} = +0.5$$

$$\text{Z in anthropology} = \frac{40 - 35}{2} = +2.5$$

In the anthropology test, Elmer was 2 1/2 standard deviations above the class mean. In his psychology test, he was only 1/2 a standard deviation above the mean. Clearly, Elmer's performance, compared to his classmates', was far superior on the anthropology test.

Consider another example. Elmer got 80% in a math test and 80% in a chemistry test. The class average was 60% on both tests. It would be tempting to conclude that Elmer performed equally well on both tests, but we know better. The data on each test were as follows:

Math Test	Chemistry Test
$\mu = 60\%$	$\mu = 60\%$
$\sigma = 15$	$\sigma = 5$
Elmer's score-80%	80%

These two distributions might look like the following:

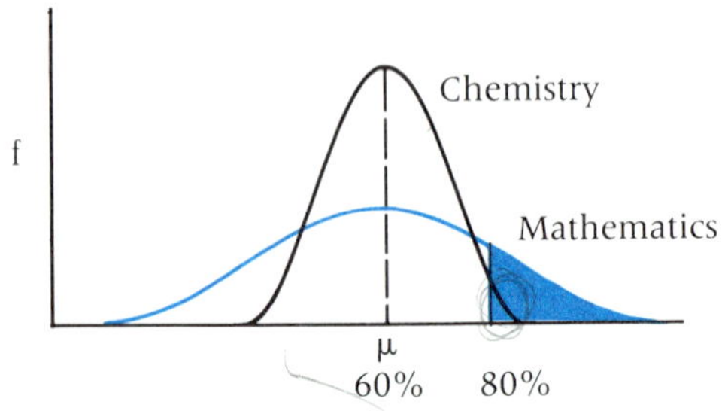

As you can see, the standard deviations differ considerably. In the math exam several students got 80% and higher but in the chemistry test very few students did as well as Elmer. Converting Elmer's grades to Z-scores we find:

$$Z \text{ (in math)} = \frac{80 - 60}{15} = +1.33$$

$$Z \text{ (in chem)} = \frac{80 - 60}{5} = +4.00$$

Elmer's score of 80% on the chemistry test was 4 standard deviations above the mean; his score on the math test was 1 1/3 units above the mean. It is necessary to know both the central tendency and the amount of spread in a distribution to determine relative standing.

SUMMARY OF TERMS AND FORMULAS

The measures discussed in this chapter are useful for determining the **location** of a score in a distribution.

The **percentile** is a score value at or below which a certain percentage of the scores fall. The **percentile rank** is a value indicated the percentage of scores at or below a particular score value.

The **Z-score** indicates the distance, above or below the mean of a distribution, that a particular scores falls in units of standard deviation

Measure	**Formula**
Percentile	$P_{PR} = L + \dfrac{\left[N\left(\dfrac{PR}{100}\right) - f_b\right]i}{f_w}$
Percentile Rank	$PR_x = \dfrac{\left[f_w\dfrac{(X - L)}{i} + f_b\right]100}{N}$
Z-Score	$Z = \dfrac{X - \mu}{\sigma}$

refer to pg 56

EXERCISES

1. Determine the 50th, 75th, and 90th percentile for the following frequency distribution.

Score Interval	f
135–139	3
130–134	12
125–129	9
120–124	10
115–119	15
110–114	7
105–109	2
100–104	2

What is the percentile rank of 117?

2. John got a 45 on his spelling test. If the class mean was 57 and the standard deviation was 6, what was John's Z-score?

3. For the following data, determine the Z-scores for each raw score.

DATA: 4, 6, 8, 9, 11, 3

4. Maria got 87% on her grammar test and 76% on her math test. Using the following data determine if she did better on her grammar test.

Grammar test	mean = 85	standard deviation = 2
Math test	mean = 70	standard deviation = 4

5. It is possible for a score to have a percentile rank of 65 and a negative Z-score. What kind of distribution must this score come from?

6. The percentile rank of a score of 89% is 62. Describe this in words.

7. Dan got 80% on an English test for which the mean was 70. He got 70% on a French test for which the mean was 80. The standard deviation for both tests was 10 and the median for both was 65. Which of the following is undeniably true?

a) the English test was easier
b) the two distributions differ in skew direction so we can't make comparisons
c) Dan's absolute Z-score for both tests is 1.00. Therefore his relative position in the two classes is the same
d) none of the above

8. A distribution of scores on a statistics test has a mean of 54 and a standard deviation of 3. What raw score would you need to be three standard deviations above the class mean?

9. A student received a Z-score of +1.30 on her history exam. If the class mean was 64% and the variance was 9, what was her raw score on the test?

CHAPTER 6

Introduction to Inference: The Normal Curve

The **normal curve** is a bell-shaped relative frequency distribution. This curve is a very useful model for many statistical procedures. Many variables, including height, weight, and IQ distribute themselves according to this bell-shaped curve. A normally distributed variable has few observations at either extreme end; the bulk of the observations lie in the middle.

Empirical vs Theoretical Frequency Distributions

> **Empirical:** Based or founded on real data or observations.
>
> **Theoretical:** Based on theory or hypothetical observations.

Empirical means "data-based." Empirical frequency distributions contain actual observations and the distribution has a fixed size. **Theoretical** distributions, on the other hand, are composed of theoretical observations and may have an infinite number of terms. The normal curve is a theoretical distribution.

The Normal Curve

> **Standard normal curve:** Theoretical bell-shaped distribution. Often simply called normal curve.

The normal curve is not a single distribution but rather a family of curves. Like the mathematical function of a circle describes a family of circles, the mathematical function of the normal curve describes a family of curves.

Many statistical procedures rely on the normal curve and, therefore, the curve has been standardized and should, more properly, be called the **standard normal curve**. The standard normal curve has Z-scores along the abscissa and relative frequency along the ordinate. In this form, all normal curves are identical.

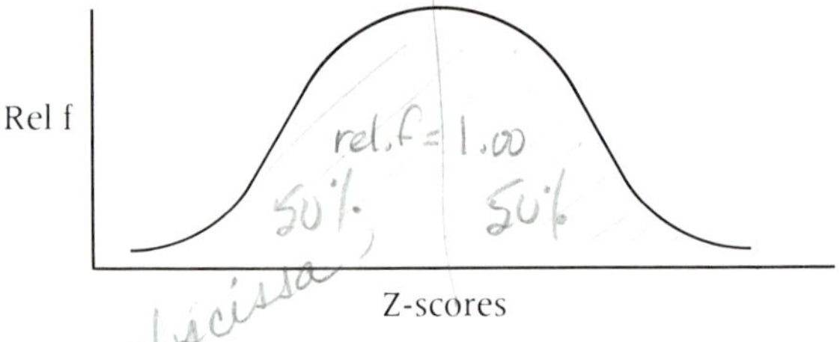

The Equation for the Standard Normal Curve

The standard normal curve is defined by a mathematical function. It is included here for your interest.

$$Y = \frac{1}{\sqrt{2\pi\sigma^2}} e^{-(X-\mu)^2/2\sigma^2}$$

Y is the height of curve at point X
X is any point on abscissa
μ is the mean
σ^2 is the variance
π is a constant = 3.1416...
e is the base of Napierian logarithms = 2.71828

This equation produces a curve with several unchanging properties.

Properties of the Standard Normal Curve

The standard normal curve has the following properties.

1. It is symmetrical.

2. It is unimodal; the mean, median, and the mode are equal.
3. The values along the abscissa are continuous.
4. The curve is asymptotic to the abscissa; because there are an infinite number of terms in the distribution, the curve never touches the abscissa.
5. The mean is zero.
6. The variance and standard deviation equal 1.
7. The area under the curve has a relative frequency or 1.00.

The entire area under the curve contains 100% or 1.00 of the frequency. Since the curve is symmetrical, 0.50 of the area is on each side of the mean. Because these properties don't change, we can solve problems with the standard normal curve.

Using the Normal Curve to Solve Problems

Table 1 at the back of this book contains two columns that provide exact areas under the standard normal curve. The first column provides areas between the mean of the curve and any particular Z-score. The second column provides areas beyond a particular Z-score. Only positive Z-scores are given because the curve is symmetrical. The area beyond a Z of +1.00, for example, is the same as the area beyond a Z of −1.00.

Finding Areas Under the Curve

If we are given a particular Z-score, we can use Table 1 to find out how much of the area lies between it and the mean; and how much of the total area lies beyond that score. When solving problems with the normal curve, first draw a rough picture. This will help you see exactly what you are doing. Ready for some examples?

How much of the total area under the curve lies between the mean and a Z-score of +1.05? We will solve these problems in steps.

Step 1. Draw a curve and shade in the area of concern.

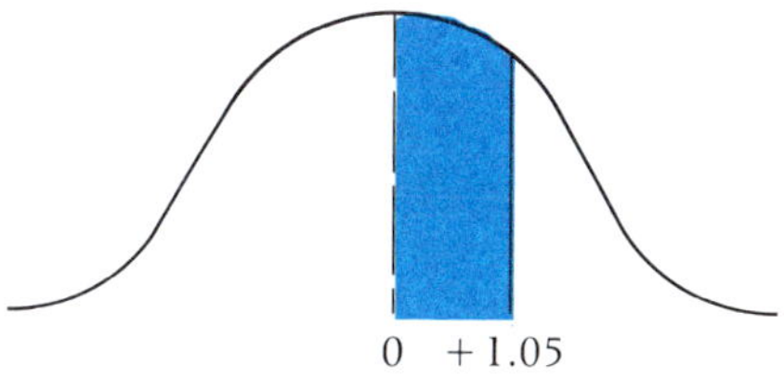

Step 2. Go to Table 1 and look up the area between the mean and a Z-score of + 1.05. Use the first column of the table.

Area = 0.3531

Approximately 35% of the total area is found between a Z of 0 (the mean) and a Z of + 1.05.

How much of the total area under the curve is found between the mean and a Z-score of − 1.74?

Step 1. Draw a curve.

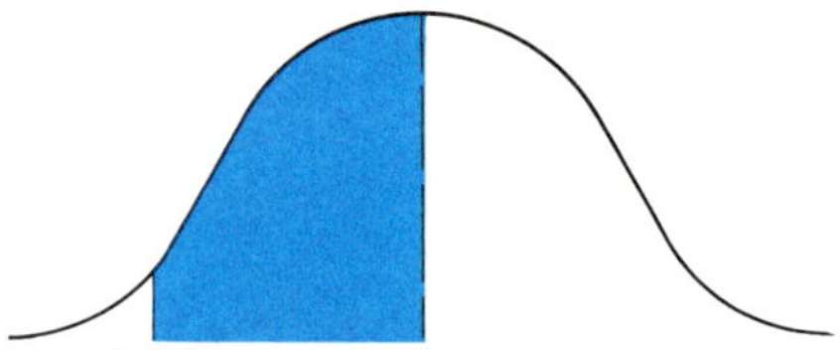

Step 2. Look up the area. Because the curve is symmetrical, we can ignore the minus sign.

Area = 0.4591

Approximately 46% of the total area is found between a Z of − 1.74 and the mean of the distribution.

How much of the area is found beyond a Z-score of 3.24?

Step 1.

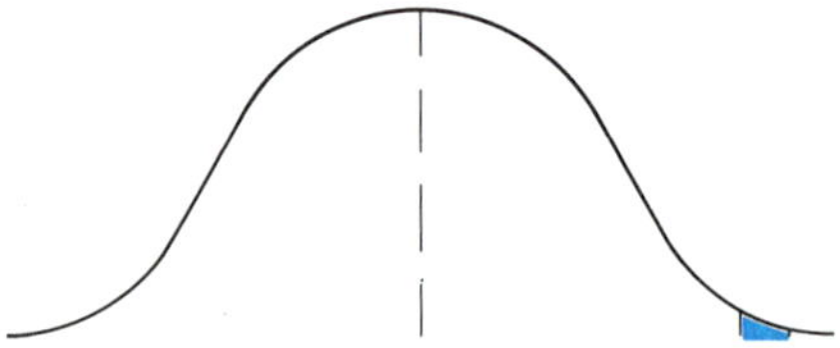

Step 2. Now we need to use the second column of the table. Look up the area beyond a Z of 3.24.

Area = 0.0006

As you can see, very little of the area is found beyond a Z of 3.24. How much of the area is found between the Z-scores of − 1.74 and + 3.24? This problem illustrates the importance of drawing a curve first.

Step 1.

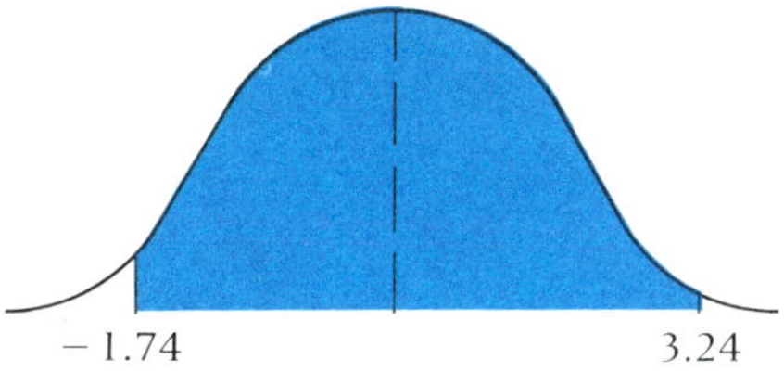

Step 2. We have already determined that 0.4591 of the area lies between a Z of −1.74 and the mean. Look up the area between the mean and a Z of 3.24. There is 0.4994 of the area between these points.

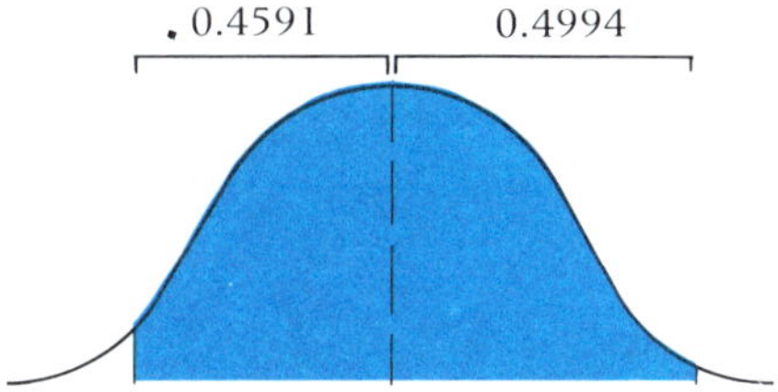

We can add these two areas together to solve our problem. There is 0.9585 or about 96% of the total area between these two scores.

As long as you have specific Z-scores, you can always determine exact proportions of the total area under the curve.

Finding Z-Scores with the Curve

We can also solve problems when the Z-score, not the area, is the unknown. For example, what Z-score has 0.1915 of the total area between it and the mean? Our step-by-step procedure is as follows:

Step 1. Draw the curve and shade in the area of concern.

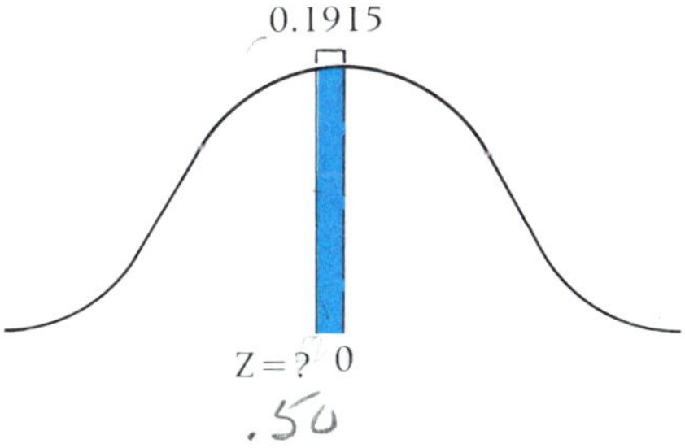

I have shaded the area below the mean. This is arbitrary. There are, of course, two Z-scores, one positive and one negative, which will solve our problem.

Step 2. Look in the first column of the table for the area of 0.1915. Look across to the left for the Z value. Either the positive or the negative value is appropriate, since both solve the problem.

$$Z\text{'s are } + \text{ or } -0.50$$

There is 0.1915 of the total area between either Z value and the mean.

Let's find the two Z-scores which cut off the middle 95% of the area from the extreme 5%.

Step 1.

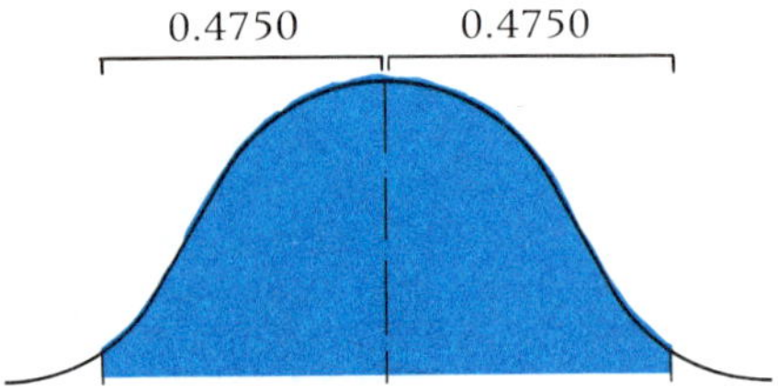

To separate the middle 95% from the extreme 5%, we need half of 95% (0.4750) on either side of the mean.

Step 2. Look up the Z which has 0.4750 between it and the mean. Z = 1.96. Since the curve is symmetrical, we know that 95% of the total area is between the Z-scores of −1.96 and +1.96.

Because the standard normal curve has unchanging properties, we can always determine exact areas when we have Z-scores, and we can find Z-scores when we have exact areas. Whenever we have a real distribution, which is normal in shape, we can use these procedures.

Solving Problems with Normal Distributions

If we know that a particular distribution of scores is normal in shape, we can use the standard normal curve. We simply translate our data into Z-score form and then use the normal curve table to solve problems about our specific distribution.

Imagine that we have a distribution of IQ scores, which is normal in shape, with a mean of 100 and a standard deviation of 15. First, let's find the proportion of scores or area below an IQ of 95.

Step 1. Draw the curve.

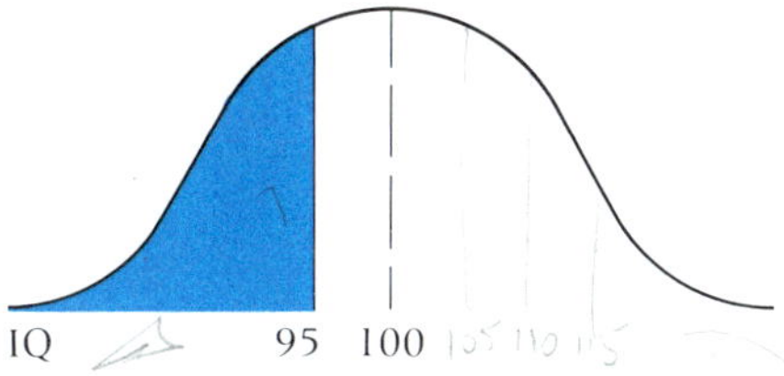

Step 2. Find the Z-score. We must transform our raw score into a Z-score so that we can use the table. Recall that the formula for converting a raw score to a Z-score is

$$Z = \frac{X - \mu}{\sigma}$$

Our Z-score is $$Z = \frac{95 - 100}{15} = -0.33$$

Step 3. Look up the area beyond a Z of -0.33.

Area = 0.3707

Approximately 37% of the distribution of IQ scores is below an IQ of 95.

What proportion of the IQ scores lies between 95 and 115?

Step 1.

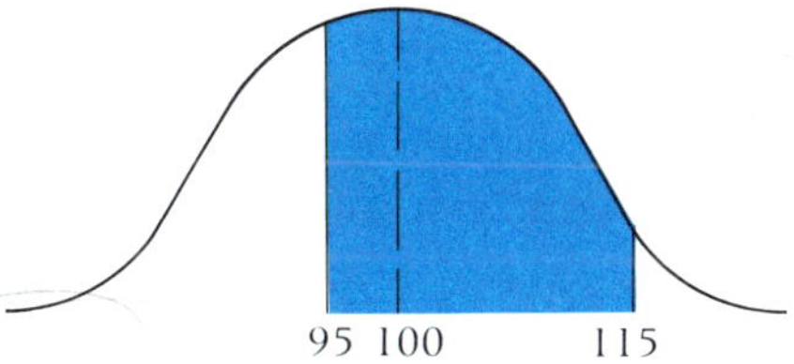

Step 2. We already know that 0.3707 of the area lies below 95, so we can just subtract this from 0.50 to find the area between 95 and the mean of 100. This is 0.1293. Next, we need to translate the score of 115 into a Z-score.

$$Z \text{ of } 115 = \frac{115 - 100}{15} = 100$$

Step 3. Looking up the area between the mean and a Z of 1.00, we find it is 0.3413.

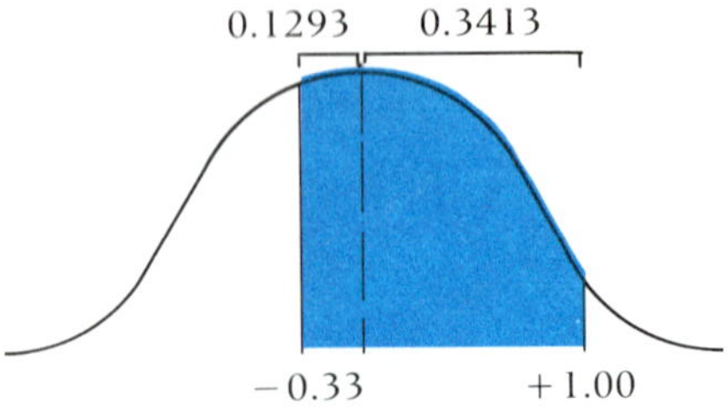

We add these two areas together and we find that 0.4706 of the area is found between the scores of 95 and 115.

Now, let's find the proportion of IQ scores between 115 and 130.

Step 1.

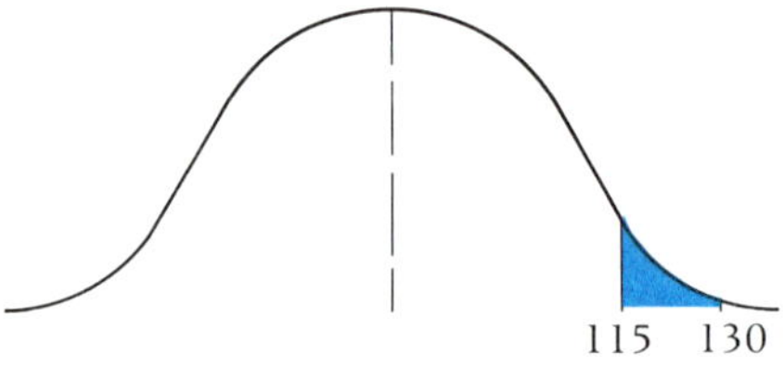

Step 2

$$\text{Z of 115 is } +1.00$$

$$\text{Z of 130} = \frac{130 - 100}{15} = +2.00$$

Step 3. There are several ways to solve this problem. Let's look up the area between the mean and a Z of +2.00. Then we subtract the area between the mean and a Z of +1.00 from that value.

Area between mean and Z = 2.00 is 0.4772
Area between mean and Z = 1.00 is 0.3413

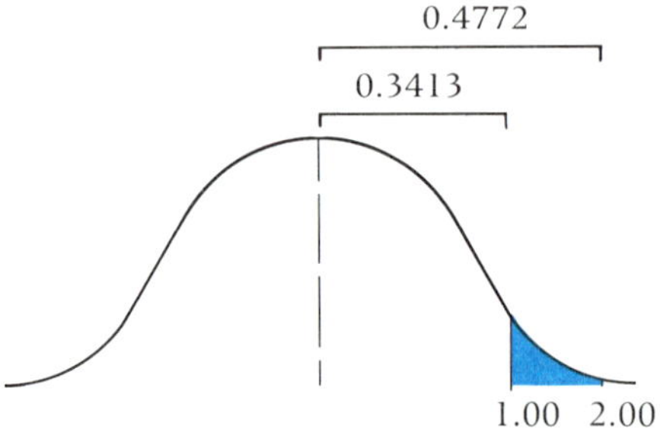

By subtraction, we find 0.1359 (0.4772 − 0.3413) of the area is between the Z's of 1.00 and 2.00. Approximately 14% of the scores lies between IQs of 115 and 130.

To continue, let's find the two IQ scores which separate the middle 90% of the distribution from the extreme 10%. We have the area and must find the scores.

Step 1.

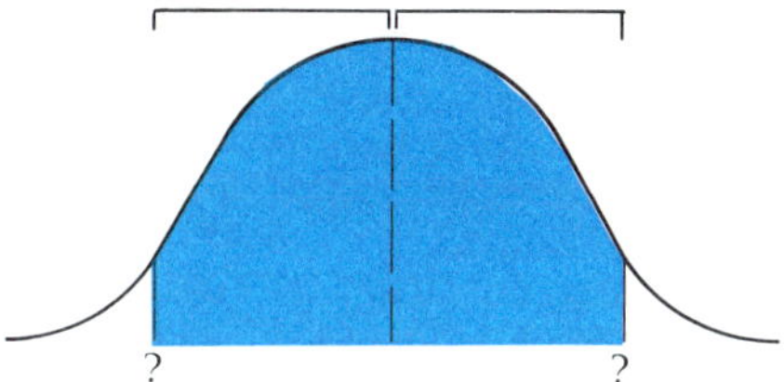

Step 2. Look up the Z score which has 0.4500 of the area between it and the mean.

$$Z = + \text{ or } - 1.645$$

We have the two Z values and we now convert them to the raw scores.

Step 3. Rearrange the Z formula so that X is the unknown.

$$X = \mu \pm Z\sigma$$
$$= 100 \pm 1.645(15)$$
$$= 124.75 \text{ and } 75.25$$

We find that 90% of the scores lies between the values of 124.75 and 75.25 and 10% lies outside these two values.

What two IQ scores cut off the middle 95% from the extreme 5%?

Step 1.

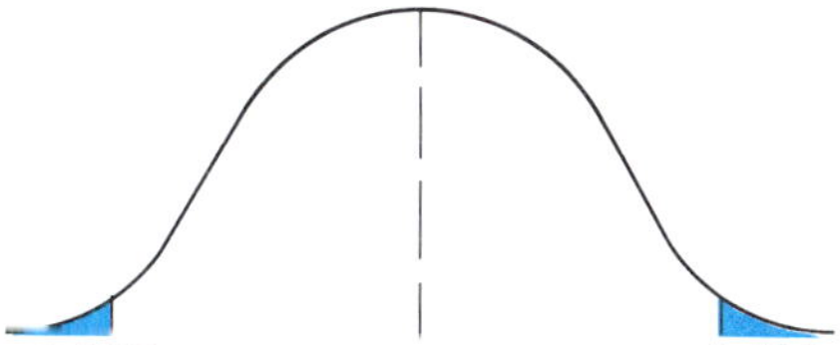

Step 2. $Z = \pm 1.96$

Step 3. $X = 100 \pm 1.96(15) = 129.40 \text{ and } 70.60$

Between IQs of 129.40 and 70.60, we find 95% of the distribution.

Whenever we have a distribution of scores that we know is normally distributed, we can use the standard normal curve as a model to solve problems about our distribution. We only need the mean and standard deviation of our distribution to use the normal curve table. Many "real" distributions are similar in shape to the normal curve, and so the normal curve table is a very useful device for solving a variety of problems.

SUMMARY OF TERMS AND CONCEPTS

The normal curve is an extremely useful tool for many statistical procedures because it has certain unchanging properties.

The **standard normal curve** is a **symmetrical** distribution with a mean, median and mode of zero. The observations along the abscissa of the curve are **infinite** and **continuous Z-scores.**

Because the area is **relative frequency** and equal to 1.00, the curve can be used as a model for solving a variety of problems about normally distributed variables.

EXERCISES

1. Determine the area in a normal distribution between $Z = \pm 2.50$.

2. Determine the area in a normal distribution between $Z = \pm 1.87$.

3. Determine the area in a normal distribution greater than $Z = 1.96$.

4. Find the two Z-scores which separate the middle .99 of the area from the extreme .01.

5. Find the Z-score below which 0.90 of the area lies.

6. Determine the percentile rank of each of the following Z-scores. (hint: area under the curve can be converted to percentage.)

a) -2.58
b) -1.96
c) 0
d) 1.64
e) 2.33

7. Assume you have a normal distribution of 5000 test scores with $\mu = 72$ and $\sigma = 12$.

a) What percentage of the scores are greater than a score of 80?
b) What percentage of the scores are below 66?
c) Between what two test scores do the middle 60% of the scores lie?
d) Beyond what score does the lower 5% of the scores fall?
e) How many scores lie below the score in d?
f) What score is exceeded by only 4 students (round to the nearest whole number)?
g) What test score is at the 25th percentile? (hint: what exactly is a percentile, what does it tell you?)

CHAPTER 7

Introduction to Inference: Probability

Assume for the moment that I am going to toss a coin and bet with you about the outcome. If I bet on heads, how many heads in a row would I have to toss before you started getting suspicious? Let's say I toss the coin three times and all three times it comes up heads. Would you accuse me of trickery? How about five heads in a row? Now would you start to doubt my honesty? What if I tossed the coin ten times and it came up heads each time? What would you have to say about that? Probably most people would be quite doubtful that a fair coin would turn up heads ten times in a row. Most people would think that this is too much for coincidence. What they are really deciding is that it is highly *improbable* that a fair coin would turn up heads as many as ten times in a row *by chance alone*. A betting person would probably conclude that the coin was not fair at all! Now, it is possible that a fair coin could behave this way, but if I had money on it, I think I would be better off rejecting such a hypothesis. It seems much more reasonable to conclude that something fishy is going on.

This simple example demonstrates the underlying logic of all statistical inference. Statistical inference involves determining the probability that a particular outcome (e.g., ten heads in a row) occurred by chance alone given a specific hypothesis about the nature of things (e.g., a fair coin was used). To understand statistical inference thoroughly, it is essential to have an understanding of basic probability.

Simple Probability

The probability of a particular **event** or **outcome** occurring is the number of ways that that event can occur, divided by the total number of possible outcomes that could occur in a situation where all outcomes are equally likely.

p(A)	=	$\frac{\#A}{\#O}$
probability of A occurring	=	$\frac{\text{number of ways A can occur}}{\text{total number of outcomes}}$

In a single toss of a coin, the probability of getting a head is the number of ways a head can occur (1) over the total number of outcomes possible (2).

$$p(H) = \frac{1}{2}$$

In the roll of a die (one of a pair of dice) the probability of rolling a four is the number of ways a four can occur (one) over the total number of possible outcomes (six).

$$p(4) = \frac{1}{6}$$

Both of these examples have outcomes that are equally likely. In a single toss of a fair coin, a head is as likely to occur as a tail. With a fair die, the likelihood of rolling a four is the same as for a 1, 2, 3, 5 or 6.

What would be the probability of obtaining an even number on a single roll of a fair die? Using the same approach we would determine the number of ways that an even number could occur over the total number of outcomes. There are three ways an even number could show up (i.e., 2, 4, 6) and the total number of outcomes remains the same (six).

$$p(\text{even number}) = \frac{3}{6}$$

If we rolled a die many times then, we would expect to get an even number half the time on the average since the probability on a single roll is one half.

Conditional Probability

When the probability of a given event is *dependent upon* or *influenced by* the occurrence or non-occurrence of a previous event, it is called **conditional probability**. The likelihood of a given event occurring is conditional upon what went on before.

Let's determine the probability of drawing a jack from a well-shuffled deck of playing cards. Since all cards are equally likely to be drawn we can determine the probability of drawing a jack by using our simple probability formula.

$$p(J) = \frac{\text{\# of ways a jack can be drawn}}{\text{total number of outcomes or cards}}$$

There are four jacks in a deck of 52 cards and so,

$$p(J) = \frac{4}{52}$$

Now let's draw a second card from our deck without putting the first one back. What is the probability of drawing another jack if the first card drawn was a jack? Since there are only three jacks left, the number of ways we can draw a second jack is 3 and the total number of cards is only 51.

$$p(\text{jack given the first card was a jack}) = \frac{3}{51}$$

This is an example of conditional probability since the outcome (jack on second draw) is conditional upon or dependent upon what happened in the first draw. The general formula for this kind of probability is

$$p(B/A) = \frac{\text{\# ways B can occur given A has occurred}}{\text{total number of outcomes}}$$

In our example: $p(J/J) = \frac{3}{51}$

> **Dependent events:** Occurrence of one alters the probability of occurrence of the other.
> **Independent events:** Occurrence of one has no effect on the probability of occurrence of the other.

What if a jack was not drawn the first time? The probability of getting a jack on the second draw, given a jack was not drawn on the first draw, would be the number of ways to draw a jack (four, since all four are still in the deck) over the total number of outcomes (51 since one card was already drawn and not replaced.)

$$\text{p(jack/non-jack)} = \frac{4}{51}$$

Dependent events are events where the occurrence of one alters the probability of occurrence of the other. Conditional probability involves dependent events. Conversely, **independent events** are those where the occurrence of one does not affect the probability of occurrence of the other. In our example, had we returned the first card drawn to the deck, the probability of drawing a jack on the second draw would have been unaffected by the outcome of the first draw. In this case, the probability of drawing a second jack would be a matter of simple probability, the number of ways of drawing a jack (four) over the total number of outcomes (52).

$$\text{p(jack on second draw given the first drawn jack was replaced)} = \frac{4}{52}$$

These two events are independent because the outcome of the second draw is not influenced by what happened on the previous draw. Conditonal probability refers only to dependent events and by replacing our first drawn card we have made the second draw independent of the first and, therefore, not a conditional event. We could describe this situation of independent events using our formula. When events are independent than the probability of event B occurring, given A has already occurred, is the same as the probability of B.

If $p(B/A) = p(B)$ then the events are **independent**.

This underlies the "gambler's fallacy." Many gambling games (e.g., roulette) involve independent events. The fact that the ball has just stopped on the 34 has absolutely no effect on its next stopping place. I am, of course, assuming a fair roulette wheel, which may be naive! Unfortunately, many gamblers think these events are dependent, and they have a multitude of schemes for trying to determine probabilities. It's really very simple. If a roulette wheel has 36 numbers the probability of the ball stopping on the number 34 is 1/36 (i.e. the number of ways a 34 can occur, 1, over the number of possible outcomes, 36), no matter what numbers have previously occurred. I should mention that this knowledge has never stopped me from betting on my favorite number 7!

Probability of Compound Events

Probability of A and B

Many situations involve questions about the probability of two or more specified events occurring rather than just a single event. We may wonder what the probability might be of obtaining a jack on the first draw and another jack on the second draw from our deck of 52 cards. The general formula for the probability of two specified events both occurring is

$p(A \text{ and } B) = p(A) \cdot p(B/A)$ when the events are **dependent**.

The probability of both A and B occurring is the product of the probability of A and the probability of B given A has occurred. We use a dot to indicate multiplication because large brackets are somewhat messy looking. We can now determine the probability of getting two jacks in a row drawn without replacing the first one from a deck of 52 cards:

$$p(J \text{ and } J) = p(J) \cdot p(J/J)$$
$$p(J \text{ and } J) = \frac{4}{52} \cdot \frac{3}{51} = 0.0045$$

On the first draw, the probability of a jack is simply the number of ways a jack can occur (4) over the total number of outcomes (52). On the second draw there are only three ways to get a second jack out of a total of 51 cards because the first jack was not replaced.

If we had replaced the first jack, making the two events independent, then the probability of the second jack would not be affected by the first draw and the probability of getting two jacks in a row would be

$$p(J \text{ and } J) = p(J) \cdot p(J)$$
$$p(J \text{ and } J) = \frac{4}{52} \cdot \frac{4}{52} = 0.0059$$

The general formula for the probability of two independent events occurring is

$$p(\text{A and B}) = p(\text{A}) \cdot p(\text{B}) \text{ when events are } \textbf{independent}.$$

This procedure can be used for more than two events as well. If all events are independent then the probability of all of them occurring is the product of their individual probabilities. Recall the coin toss experiment from the beginning of this chapter. I wondered how many heads in a row I could toss before you became suspicious about the fairness of my coin, and suggested that 10 heads in a row would cause most people to doubt me. We can now determine what the probability really is of obtaining 10 heads in a row with 10 tosses of a fair coin. This is just the probability of getting a head *and* a head *and* a head etc.

$$p\begin{bmatrix}\text{10 heads}\\ \text{in a row}\end{bmatrix} = p(H) \cdot p(H) \cdot p(H) \cdot p(H) \cdot p(H) \cdot p(H) \cdot p(H) \cdot p(H) \cdot p(H) \cdot p(H)$$

$$= \frac{1}{2} \cdot \frac{1}{2} \cdot \frac{1}{2} \cdot \frac{1}{2} \cdot \frac{1}{2} \cdot \frac{1}{2} \cdot \frac{1}{2} \cdot \frac{1}{2} \cdot \frac{1}{2} \cdot \frac{1}{2}$$

$$= \frac{1}{1024} = 0.00098$$

As you can see this is a mighty unlikely series of events!

Probability of A or B

We have seen how to determine the probability of two (or more) events occurring together. We can also ask questions about the probability of either of two events occurring. This is called the probability of **A** or **B** and the general formula is

$$p(\text{A or B}) = p(\text{A}) + p(\text{B}) - p(\text{A and B}) \quad \text{General Formula}$$

Let's go back to our deck of playing cards. What would be the probability of drawing a jack or a red card in a single draw from our deck of 52? Half of the deck (26) is red and two of the cards are black jacks, so the number of ways to get a jack or a red card must be 28 out of the total number of outcomes (52). Let us use our general formula to verify this.

> **Mutually exclusive events:** Two events that cannot occur at the same time.

$$p(\text{J or R}) = p(\text{J}) + p(\text{R}) - p(\text{R and J})$$
$$= \frac{4}{52} + \frac{26}{52} - \frac{2}{52} = \frac{28}{52} = 0.54$$

This formula is always used when two events can *occur together.* We needed to subtract the p(R and J) so that we didn't count the red jacks twice. A red card can also be a jack so these events can occur together. When two events cannot occur at the same time, they are said to be **mutually exclusive**. Drawing a jack or a queen, for example, would be mutually exclusive events since you cannot draw a jack and a queen in one draw. If one event has occurred, the other cannot have occurred.

In the case of mutually exclusive events, the last term in the above formula is zero, since A and B cannot occur at the same time. The probability of drawing a jack or a queen in a single draw is simply the sum of their individual probabilities.

$$p(\text{J or Q}) = p(\text{J}) + p(\text{Q})$$
$$= \frac{4}{52} + \frac{4}{52} = 0.1538$$

The general formula, then, for determining the probability of one event or the other event occurring when events are mutually exclusive is:

> $p(\text{A or B}) = p(\text{A}) + p(\text{B})$ for **mutually exclusive events.**

The examples we have looked at have been quite simple. We began by learning that the probability of occurrence of a given event (A) is found by dividing the number of ways A can occur by the total number of outcomes possible. The total number of outcomes and the number of ways A can occur are easy to determine when you are dealing with dice, coins, or a couple of draws from a deck of playing cards. What if I ask you to determine the probability of being dealt a royal straight flush (i.e., A, K, Q, J, 10 all in the same suit) from a deck of 52 cards? Could you do it?

It's not too tricky to figure out how many ways you could get a royal straight flush. There are four suits, so there are only four ways, one for each suit. However, determining the total number of 5-card hands possible would daunt most of us! Luckily, we have methods for doing just this sort of thing without the necessity of enumerating every possibility.

Methods of Counting

Permutations

A **permutation** is an **ordered sequence** of things, objects or events. A different order of the same objects is a different permutation of those objects. Consider a relay swim team whose members are Hewey, Dewey, and Susan. There are several ways we could arrange our three swimmers in terms of position.

	Teams					
	1	2	3	4	5	6
First swimmer	Hewey	Hewey	Dewey	Dewey	Susan	Susan
Second swimmer	Dewey	Susan	Hewey	Susan	Hewey	Dewey
Third swimmer	Susan	Dewey	Susan	Hewey	Dewey	Hewey

There are six different arrangements that could be made with our three swimmers. We are simply arranging our three objects (swimmers) into different orders or permutations.

What would happen if we added one more swimmer (Lewey)? We would end up with a whole lot more possibilities, 24 to be exact. As you can imagine it would get quite tedious to delineate all possible arrangements. Fortunately, we have a formula to do this work for us. The formula is

$$_nP_r = \frac{n!}{(n - r)!} \text{ for } \textbf{ordered sequences}$$

We will use the letter **n** to stand for the total number of objects we have to work with, and the letter **r** to stand for the number of objects we need in a single arrangement. The number of ordered sequences of r objects that can be selected from a total of n objects is symbolized by $_n\mathbf{P}_r$. $_nP_r$ can be read as the number of permutations of n objects taken r at a time.

The exclamation mark means **factorial**, which I have always thought was a somewhat mysterious operation. Here is how it is done.

$$n! = n(n-1)(n-2)(n-3)\ldots 1$$
$$\text{e.g., } 6! = 6(5)(4)(3)(2)(1) = 720$$

At this point it is important to mention that *0! is always 1.* Just accept this on faith.

To illustrate this, let's select four objects which we will cleverly call A, B, C and D. We can determine how many different ordered sequences of two objects can be selected from our total of four in the following way:

Objects	Sequences of 2 at a time		
A, B, C, D	A B,	A C,	A D
	B A,	B C,	B D
	C A,	C B,	C D
	D A,	D B,	D C

Note that AB and BA are *different permutations.* The different order of objects makes them different permutations.

By counting up our sequences, we can see that there are 12 permutations of 4 objects taken 2 at a time. We can use our formula to verify this.

$${}_4P_2 = \frac{4!}{(4-2)!} = \frac{4 \cdot 3 \cdot 2 \cdot 1}{2 \cdot 1} = 12$$

Let us look at a more useful example. Those of you interested in horse racing may find this profitable. What is the probability of selecting the first, second, and third in a race involving a field of 10 horses? In horse-racing terminology this is called picking the win, place, and show horse. Our general approach tells us that we must divide the number of ways we could pick first, second, and third by the total number of outcomes in such a race. Clearly, there is only one way of selecting the first three horses in the correct position. However, the total number of possible sequences of three horses from a field of 10 is not so obvious. We can use our formula to solve this problem.

$$\text{total number of outcomes} = {}_{10}P_3 = \frac{10!}{(10-3)!} = 720$$

There are 720 ordered sequences of 10 horses selected in threes. The probability, then, of picking the winning three horses in the correct order can now be determined.

$$\text{p (picking win, place and show)} = \frac{\text{\# ways to pick 1st, 2nd and 3rd}}{\text{total \# of outcomes}}$$
$$= \frac{1}{720} = 0.00139$$

You might want to consider this the next time you gamble on a Trifecta! Of course, it is important to point out that the above example assumes we are randomly choosing the three horses from the field. Most of us, I am sure, never behave so frivolously when it comes to betting on horse races. We use important information, such as the attractiveness of the horses' names and their colour, to determine our choices!

Combinations

A **combination** is a **set** of things, objects or events. Order is not important. *AB and BA is the same combination.* The formula to determine the number of combination of n objects taken r at a time is

$$_{n}C_{r} = \frac{n!}{(n - r)!r!} \quad \text{for } \textbf{unordered sequences}$$

Using our four objects (A, B, C, and D) we can determine the number of combinations of two objects which can be selected.

Objects	Combinations of 2
A, B, C, D	AB, AC, AD
	BC, BD, CD

There are only *six combinations* but there were *twelve permutations*. For every one combination, then, there were two permutations. **In any situation, there will be r! permutations for every one combination**. This is why you find r! in the denominator of the combinations formula.

We can verify our example by using the combination formula for 4 objects selected 2 at a time.

$$_{4}C_{2} = \frac{4!}{(4 - 2)!2!} = \frac{4 \cdot 3 \cdot 2 \cdot 1}{(2 \cdot 1)(2 \cdot 1)} = 6$$

Again, using our four objects, let us determine the number of combinations and permutations of three objects that can be selected.

Objects	Combinations of three	Permutations of three
A, B, C, D	ABC, ABD, ACD, BCD	ABC, ABD, ACD, BCD
		ACB, ADB, ADC, BDC
		BAC, BAD, CAD, CBD
		BCA, BDA, CDA, CDB
		CBA, DBA, DCA, DBC
		CAB, DAB, DAC, DCB

We have 24 permutations but only four combinations. There are 6 permutations for every one combination. Note that r! is 6 so there are r! permutations for every one combination.

Using the formulas for the above example:

$$_4C_3 = \frac{4!}{1!3!} = \frac{4 \cdot 3 \cdot 2 \cdot 1}{1 \cdot 3 \cdot 2 \cdot 1} = 4$$

$$_4P_3 = \frac{4!}{1!} = \frac{4 \cdot 3 \cdot 2 \cdot 1}{1} = 24$$

Binomial Probability

When I was an undergraduate student taking statistics, the topic of **binomial probability** almost defeated me. I could not understand why it was important. It is important, however, because many things in this world have only two outcomes, e.g., on/off, yes/no, pregnant/not pregnant. It's really not very complicated at all.

A special case in probability exists when we are trying to determine the number of combinations when there are only two outcomes of interest for each of several trials. A variable having only two outcomes of interest is also called a **dichotomous variable**. Consider penny tossing again. You may be wondering if all statisticians spend a lot of time tossing pennies. Well... some do and some don't.

In a penny tossing experiment there are only two outcomes of interest, heads and tails. If I were to toss a penny ten times in a row what would be the probability of obtaining exactly two heads? This is a case of binomial probability. We are talking about combinations not permutations, since the question does not specify the order in which the events must occur. We are specifying that only two of the ten tosses must be heads. To determine the probability of such an event, we must first determine the probability of any one arrangement of two heads and eight tails, and then determine how many such arrangements or sequences are possible.

One sequence which fits our requirements could be the following:

H,H,T,T,T,T,T,T,T,T

What is the probability of tossing this sequence of ten coins? We learned earlier that the probability of several specified events occurring is the product of their individual probabilities, when the events are independent. The events in the above example are definitely independent; earlier events do not alter the probability of later events occurring. The probability of getting a head in a single toss of a coin is 1/2 and the probability of tossing a tail is also 1/2, so the probability of the above sequence of heads and tails is

$$p(H,H,T,T,T,T,T,T,T,T) = \frac{1}{2} \cdot \frac{1}{2} \cdot \frac{1}{2} \cdot \frac{1}{2} \cdot \frac{1}{2} \cdot \frac{1}{2} \cdot \frac{1}{2} \cdot \frac{1}{2} \cdot \frac{1}{2} \cdot \frac{1}{2} = \frac{1}{2^{10}}$$

$$= \frac{1}{1024}$$

Another sequence of events that would satisfy our requirement of exactly two heads in ten tosses is H,T,H,T,T,T,T,T,T,T. The probability of this sequence occurring is also 1/1024. In fact, the probability of *any* particular sequence of two heads and eight tails is 1/1024.

Now that we know the probability of any one sequence occurring, we need to figure out how many sequences fit our requirement of two heads and eight tails.

The number of sequences of two heads in ten tosses of a coin is simply the number of combinations of ten things taken two at a time. Using our combinations formula:

$$_{10}C_2 = \frac{10!}{(10-2)!2!} = 45$$

There are 45 different ways that we can get exactly two heads in ten tosses of a coin. Each way has a probability of occurring of 1/1024. The probability, then, of getting two heads in ten tosses of a coin is

$$p(\text{exactly 2H in 10 tosses}) = 45 \cdot \frac{1}{1024} = 0.0439$$

This may seem like a rather long and drawn out way to solve a seemingly simple problem. It is! Fortunately we have a formula for determining binomial probabilities.

In a sequence of **n** independent trials with only two outcomes of interest (we call them "success" and "failure"), where **p** is the probability of success and **q** is the probability of failure, the probability of exactly **r** successes is

$$_nC_r p^r q^{n-r} = \frac{n!}{(n-r)!r!} p^r q^{n-r} \qquad \text{Binomial Probability}$$

Success and failure are mutually exclusive events. **p** is the probability of success and **q** is the probability of failure. **q** *always equals* **1 − p**. The terms success and failure are really quite arbitrary. In general, success is the event in which you are interested. In our example above, a success is getting a head. p is the probability of getting a head in one trial (i.e., 1/2). q is the probability of not getting a head (i.e., a failure) and in our case that probability is also 1/2. The probability of a success does not have to equal the probability of a failure, their individual probabilities must add up to 1, however. (i.e., they must be mutually exclusive events).

Back to our coin tossing experiment. We have already determined the probability of getting exactly two heads in ten tosses of a coin. Now let's verify this with our formula.

$$p(2H \text{ in } 10 \text{ tosses}) = {}_{10}C_2 p^r q^{n-r}$$
$$= \frac{10!}{2!(10-2)!}\left(\frac{1}{2}\right)^2\left(\frac{1}{2}\right)^8$$
$$= 45 \cdot \left(\frac{1}{2}\right)^{10} = \frac{45}{1024} = 0.0439$$

As you can see we ended up with the same number. So now you can go ahead and determine the probability of getting exactly 13 heads with 54 tosses of a fair coin, or 43 heads with 100 tosses of a fair coin, or...

Frequency Distributions as Probability Distributions

Any frequency distribution can be viewed as a distribution of probabilities. Below are hypothetical final grades for a statistics class. The grades range from 9 through 1, where grades of 3, 2 and 1 are considered failing.

Grade	f	rel. f
8	30	0.15
7	60	0.30
6	80	0.40
5	20	0.10
4	10	0.05
	N = 200	

Viewing this distribution as a probability distribution we can see that the probability of a *randomly selected* student receiving a grade of 7 is 0.30. Further, the probability of a student receiving a grade of 6 or better would be $p(6) + p(7) + p(8) = 0.85$.

Any frequency distribution can be viewed as a probability distribution. This underlies some important procedures we will study later.

SUMMARY OF TERMS AND FORMULAS

Simple probability

$$p(A) = \frac{\#A}{\#O}$$

Given equally likely events

Conditional probability

$$p(B/A) = \frac{\#B/A \text{ has occurred}}{\#O/A \text{ has occurred}}$$

When events are dependent

Compound probability

Dependent events	$p(A \text{ and } B) = p(A) \cdot p(B/A)$
Independent events	$p(A \text{ and } B) = p(A) \cdot p(B)$
Not mutually exclusive	$p(A \text{ or } B) = p(A) + p(B) - p(A \text{ and } B)$
Mutually exclusive	$p(A \text{ or } B) = p(A) + p(B)$

Methods of Counting

Permutations $\quad {}_nP_r = \frac{n!}{(n-r)!}$

Combinations $\quad {}_nC_r = \frac{n!}{(n-r)!r!}$

Binomial Probability

$${}_nC_r p^r q^{n-r} = \frac{n!}{(n-r)!r!} p^r q^{n-r}$$

EXERCISES

1. Which of the following pairs of events are mutually exclusive?

a) rolling a six/rolling an odd number in one roll of a die
b) drawing an Ace/drawing a red card in one draw from a deck
c) flipping a head/flipping a tail in one flip of a coin
d) being pregnant/not being pregnant

2. Which of the following pairs of events are independent?

a) flipping a head on the first flip/flipping a head on the second flip of a coin
b) drawing an Ace first/drawing a red card second in two draws from a deck without replacement

c) drawing an Ace first/drawing a red card second in two draws from a deck with replacement

d) being accepted by the University of Alberta/being accepted by the University of Ottawa

3. Imagine that the probability of your getting married in the next 5 years is 0.80. The probability that you will be a parent within the next 5 years is 0.70.

a) What is the probability that you will be a parent and be married?

b) What are your chances of either being married or being a parent?

4. The probability that a certain (unenlightened) company will hire a woman to be chief executive is 0.15. The probability that this same company will hire an unmarried person is 0.25. What is the probability that this company will hire:

a) A single woman

b) A single man

c) either a woman or a married person

d) a married woman

e) are the events in c) mutually exclusive?

5. Sheila wants to move to another city. If Sheila's chance of getting a job in Vancouver is 0.10 and her chance of getting one in Halifax is 0.25, what is the probability that Sheila will be offered both jobs? What is the probability that Sheila will get either the Vancouver position or the Halifax position?

6. My glove compartment contains 3 red fuses and 5 green fuses. If I draw 6 fuses with replacement, what is the probability of getting exactly 2 red and 4 green fuses?

7. In a horse race with 6 horses and 6 post positions, how many different orders of horses can we have?

8. How many different teams of 5 volleyballers could be made from 8 players?

9. In "Over-the-Line" baseball, there is a pitcher, batter, left fielder and right fielder. If we have 10 players, how many different teams could we have if different positions of the same players is a different team?

CHAPTER 8

Introduction to Inference: The Random Sampling Distribution

This chapter introduces the **random sampling distribution**, a distribution which plays a central role in inferential statistics. Although it may seem that we have spent a lot of time introducing statistical inference, it is important to have a thorough understanding of the basic concepts underlying inferential techniques.

Statistical Inference

> **Parameter:** A characteristic of a population.
> **Statistic:** A characteristic of a sample.

It is not always practical and often not even possible to collect all the observations in a population. Statistical inference allows us to draw conclusions about values of **population parameters** from the values of **statistics** computed from samples drawn from populations.

Descriptive vs Inferential Statistics

Descriptive statistics requires:

1. specifying the population of interest;
2. collecting all observations from the population;

3. computing summary values such as the mean, standard deviation, and
4. using these summary values to describe the properties of the population.

Technically, since we are examining the entire population, we should not use the word "statistics." Rather, we should talk about "descriptive parameters." However, it has become common to use the term "statistics" even when describing entire populations.

Inferential statistics requires:

1. obtaining a sample(s) from the population(s) of interest;
2. computing estimates of parameters from the sample data to
3. make inferences about the corresponding population parameters from which the sample was drawn.

The Random Sample

Random sample: One collected such that all members of the population are equally likely to be included.

A **random sample** is drawn so that all observations in the population have an equal likelihood of being included in the sample. The term "random sample" does not define what the sample is like. Rather, it describes *how* the members of the sample were selected. Random sampling is a procedure for collecting observations from a population. It is a common technique in inferential statistics.

Sampling Fluctuation

Two samples randomly drawn from the same population will rarely be identical. It is a fundamental fact that sample outcomes vary. This is called **sampling fluctuation** or sampling error. Imagine a population of IQ scores of all Canadians. The mean of the population is 100 and the standard deviation is 15. Let's put all these IQ scores in a hat, shake them up, and select 10 scores randomly. We compute the mean of our sample to be 95. If we used this value as our inference about the population mean, we would be wrong. Whenever we make a statistical inference *we may be wrong*!

The key is to discover what values of a statistic would occur, and with what frequency, when sampling repeatedly from the same population. If we know what outcomes are likely to occur, then we can make conclusions about a particular outcome we have found.

The Random Sampling Distribution

The **random sampling distribution** is critical in inferential statistics. It is a *relative frequency distribution* of the values of some statistic calculated for *all possible samples of a fixed size* drawn at random from a given population.

Let's return, for a moment, to the example of the biased coin. How many heads in a row could I toss before you became suspicious about the fairness of my coin? Perhaps you would start feeling doubtful after the sixth head in a row. Let's analyze this situation as a statistician might.

We obtain a coin that we know is fair, toss it six times, and record how many heads occur. If we repeat this experiment enough times, we will get six heads in a row, even though our coin is fair. However, this outcome will be quite rare. If we could determine how often six heads occur in a row with a fair coin, we would know the *probability* of such an outcome in a single experiment of six tosses. If we computed the frequency of all possible outcomes of six coin tosses, we would have a frequency distribution. We could convert this to a relative frequency distribution giving us a random sampling distribution for six tosses of a fair coin, repeated many times.

Let's say that the relative frequency of six heads in a row when the experiment is repeated an infinite number of times is $1/2^6 = 0.016$. This probability is very low. If we find ourselves in a situation where we are betting on the fairness of a coin that came up heads six times in a row, we should put our money on the biased side, not on the fair side. It is extremely unlikely that a fair coin would produce such an outcome.

Random sampling distributions describe the probabilities of outcomes in specified situations. When we know what is likely to occur in a given situation, we are able to make informed decisions about a particular outcome that has occurred.

A random sampling distribution can be constructed for any statistic. The general procedure for constructing a random sampling distribution is the following:

Step 1. Randomly select a sample(s) of observations of some fixed size from a given population.

Step 2. Calculate some statistic for that sample.

Step 3. Return the sample to the population.

Step 4. Repeat the first three steps until all possible samples have been drawn.

Step 5. Place all statistics in a relative frequency distribution.

A random sampling distribution is named after the statistic of interest. For example, if the statistic calculated on the samples was the median, we would call the relative frequency distribution "the random sampling distribution of the median." If the mode was calculated, it would be the "random sampling distribution of the mode."

In our studies, we will learn about many random sampling distributions. One very useful sampling distribution in inferential statistics is the **random sampling distribution of the mean**.

The Random Sampling Distribution of the Mean

To construct a random sampling distribution of the mean we follow these steps:

Step 1. Randomly select a sample of some fixed size from the population of interest.

Step 2. Calculate the sample mean.

Step 3. Return the sample to the population.

Step 4. Repeat steps 1–3 until all possible samples have been drawn.

Step 5. Place the sample means in a relative frequency distribution.

We now have a distribution of sample means. This random sampling distribution of the mean has certain invariable characteristics.

The Mean

Recall that the symbol "μ" stands for the mean of a population of observations. We need new notation to refer to the mean of a sample drawn from a population. We will use $\overline{\mathbf{X}}$ (Xbar). The random sampling distribution of the mean is the relative frequency distribution of the means calculated for all possible samples drawn from a population. This distribution of sample means has a mean: itself. The symbol for the mean of the sampling distribution of means is $\boldsymbol{\mu_{\overline{X}}}$.

The mean of the random sampling distribution of the mean ($\mu_{\overline{X}}$) equals the mean of the raw score population (μ). To summarize:

μ is the mean of the population.
$\overline{X}$ is the mean of the sample.
$\mu_{\overline{X}}$ is the mean of the sampling distribution of means.

$\mu_{\overline{X}} = \mu$.

The Variance

The symbol for the variance of a population of raw scores is "σ^2." The variance of the sampling distribution of means indicates the variability of the sample means around the mean of the distribution. The symbol for the variance of a random sampling distribution of means is $\boldsymbol{\sigma^2_{\bar{x}}}$. The variance of the sampling distribution of means is *not* equal to the variance of the population. The variability of the means around $\mu_{\bar{x}}$ is less than the variability of the raw scores around μ. In fact, the variance of the sampling distribution ($\sigma^2_{\bar{x}}$) will be less than that of the population by a factor of the size of the sample. We will use lower case **n** to refer to sample size. The variance of the sampling distribution is related to the variance of the population in the following way: $\boldsymbol{\sigma^2_{\bar{x}} = \sigma^2/n}$. The variance of the sampling distribution is $1/n$ smaller than the variance of the population of raw scores. Note that the variance of the sampling distribution will be less for a distribution constructed from large samples than for a distribution constructed from small samples. *As sample size increases, the variance of the distribution of means decreases*. To summarize:

σ^2 is the variance of the population.
$\sigma^2_{\bar{x}}$ is the variance of the sampling distribution of means.
$\sigma^2_{\bar{x}} = \sigma^2/n$ where n is the sample size.

The Standard Deviation

Standard error: Standard deviation of a random sampling distribution

The symbol for the standard deviation of a population is "σ." The symbol for the standard deviation of the random sampling distribution of the mean is $\boldsymbol{\sigma_{\bar{x}}}$. This measure is used so frequently, it has a special name. It is called the **standard error of the mean**. It is easier to say "standard error of the mean" than "standard deviation of the random sampling distribution of means."

Recall that the variance of the sampling distribution is related to the variance of the population by $\sigma^2_{\bar{x}} = \sigma^2/n$. You can see, then, that the standard error is related to the standard deviation of the population in the following way: $\boldsymbol{\sigma_{\bar{x}}} = \sigma/\sqrt{n}$. The standard error is smaller than the standard deviation of the population by a factor of $1/\sqrt{n}$. To summarize:

σ is the standard deviation of a population.
$\sigma_{\bar{x}}$ is the standard deviation of the sampling distribution of means (called the standard error).
$\sigma_{\bar{x}} = \sigma/\sqrt{n}$ where n is the sample size.

The Shape: The Central Limit Theorem

The **Central Limit Theorem** is likely the most important theorem in inferential statistics. It states the following:

1. **The random sampling distribution of means tends toward a normal distribution, irrespective of the shape of the original raw score population.**
2. **This tendency increases as sample size increases.**

In other words no matter what the shape of the original population of raw scores, the relative frequency distribution of means, computed for all possible samples drawn from that population, will approach the normal distribution. Furthermore, the shape will be closer to normal for larger sample sizes.

In Chapter 6, we studied the characteristics of the standard normal distribution. Those same characteristics hold for the random sampling distribution of the mean when the sample size is large enough. The normal curve may be used as a model, then, for a sampling distribution of means when:

1. The original population is normal or close to normal in shape or
2. The original population is not normal but the sample size is large.

The Central Limit theorem is important because it allows us to solve problems using the normal curve as a model.

Solving Problems with the Random Sampling Distribution of the Mean

In Chapter 6, we learned how to convert a raw score into a Z-score to solve problems with the normal curve. Recall that the formula to convert a raw score to its corresponding Z is

$$Z = \frac{X - \mu}{\sigma} \qquad \text{Z for a raw score.}$$

We were able to find the proportion of the total area under the curve between two Z-scores and to find Z-scores when we knew the area. We learned to use Table 1 to solve problems with any raw score distribution that was normal in shape.

We know that the random sampling distribution of means is normal in shape under certain conditions; we can use the same table and procedure for solving problems about means of samples that we used for solving problems about raw scores. We need only to convert the sample mean to its corresponding Z-score so that we can use the normal curve table.

Converting a sample mean to a Z-score is done in a similar manner to converting a raw score to a Z-score. The mean of the distribution is subtracted from the corresponding sample statistic and this difference is divided by the standard deviation of the distribution of the statistic. In the case of a sample mean, our formula is

$$Z = \frac{\overline{X} - \mu_{\overline{X}}}{\sigma_{\overline{X}}} \qquad \text{Z for a sample mean}$$

We need to know the value of the population mean (μ) and the value of the population standard deviation (σ). We know the mean of the sampling distribution ($\mu_{\overline{X}}$) is equal to the population mean. We can find the standard error ($\sigma_{\overline{X}}$) by dividing σ by $\sqrt{n}$. Ready for an example?

Consider a normal population of raw scores whose mean is 130 and whose standard deviation is 10. Let's randomly select a quantity of samples of 25 observations from this population and compute the mean for each. How often would we expect to get sample means between 128 and 132? We will solve this problem with the same strategy we used in Chapter 6.

$$\text{DATA: } \mu = 130$$
$$\sigma = 10$$
$$n = 25$$

Remember the following sequence.

Step 1. Draw a curve.

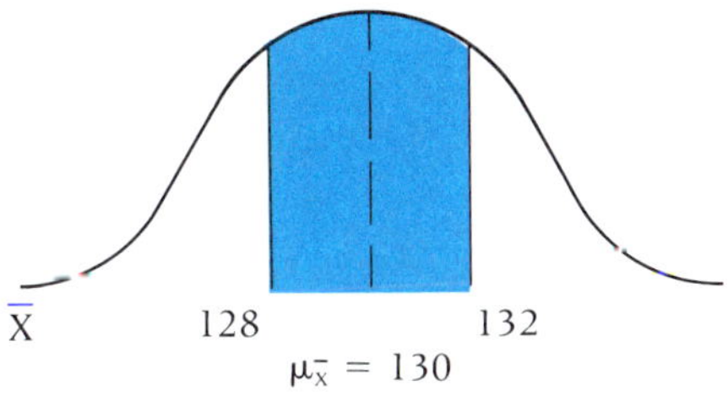

We know that the mean of the sampling distribution is 130 because it is equal to the mean of the population.

Step 2. Convert the sample means to Z-scores.

$$Z_{128} = \frac{\overline{X} - \mu_{\overline{X}}}{\sigma_{\overline{X}}} = \frac{128 - 130}{10/\sqrt{25}} = -1.00$$

$$Z_{132} = \frac{132 - 130}{2} = +1.00$$

Step 3. Look up the area between the two Z values in the normal curve table.

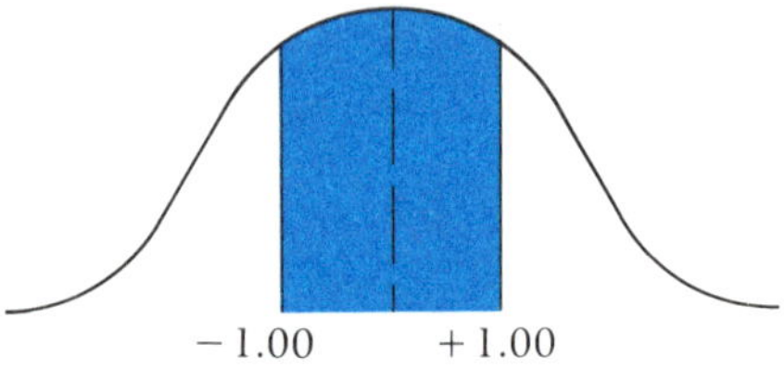

Area between the mean and a Z of +1.00 is 0.3413.
Area between the mean and a Z of −1.00 is 0.3413
Area between the two Z values is 0.6826

Therefore, when we draw many samples from this population, we would expect to get means between 128 and 132 about 68% of the time.

Let's try another example. Between what two values would we expect 99% of the means to fall?

Step 1.

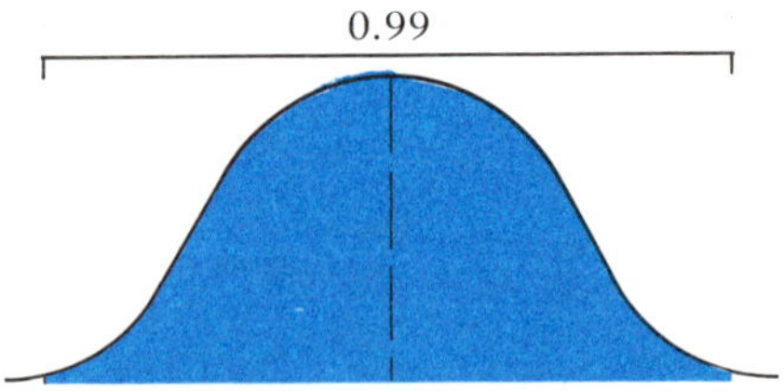

Step 2. Look up the two Z values between which .99 of the total area lies, (i.e., 0.4950 on each side of the mean).

$$Z = \pm 2.58$$

Step 3. Rearrange the formula so $\overline{X}$ is the unknown.

$$\overline{X} = \mu_{\overline{X}} \pm Z\sigma_{\overline{X}} = 130 \pm 2.58\,(10/\sqrt{25}) = 130 \pm 5.16$$
$$= 124.84 \text{ and } 135.16$$

We would expect 99% of the means to fall between 124.84 and 135.16 and the remaining 1% to fall beyond these two points.

How about a third example? How often would we expect to find sample means as large or larger than 145, drawn from our population?

Step 1.

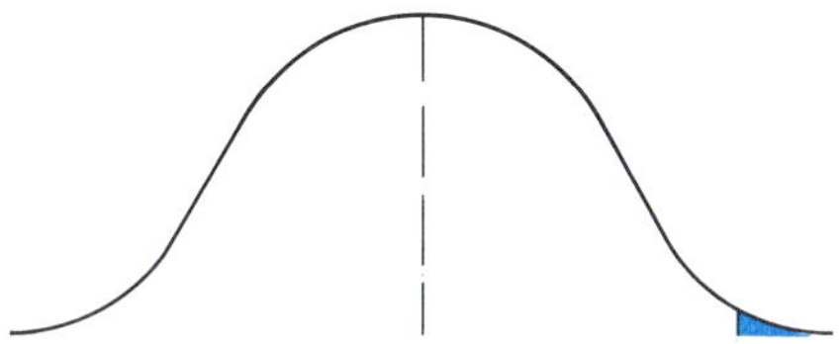

Step 2.

$$Z = \frac{145 - 130}{2} = 7.5$$

Step 3. The tables don't go this far. Out best answer is that we would never (or hardly ever) expect to get means this large. If a friend told you that she randomly selected 25 observations from a normal population whose mean is 130 and whose standard deviation is 10, and computed a sample mean of 145, your skepticism would be understandable. The probability of such an outcome, given that the information about the population is true, is incredibly low.

However, what if your friend could prove that her calculation of the mean was correct and that she did, in fact, randomly select 25 observations? What can you conclude? Two possibilities come to mind. Either she obtained an amazingly rare outcome or her information about the population from which she drew her sample is wrong.

Remember the coin toss experiment where 6 heads occurred in a row? Two explanations again come to mind. Either an extremely rare outcome occurred with a fair coin or the coin was not a fair one. In the latter explanation, expectations about how the coin should behave were wrong. This describes the logic of many inferential techniques. Inferences are based on the probability of a particular sample outcome occurring when certain hypotheses are made about the population from which the sample was selected.

The random sampling distribution of the mean is very important in statistical inference. Knowing its properties allows us to make decisions about outcomes of experiments. Another important random sampling distribution involves the differences between means of two populations.

The Random Sampling Distribution of the Difference Between Means

The **random sampling distribution of the difference between means** is a relative frequency distribution of the differences between sample means drawn from two populations with equal means. If we draw samples from populations of raw scores having equal means, we would not expect the sample means to be identical. Remember sampling fluctuation? The question we must ask is "How different would we expect sample means to be, when the samples are drawn from populations with equal means?" To answer this question, we need to determine the kinds of differences that would occur. We must construct a random sampling distribution of differences between pairs of means drawn from two populations with equal means. This distribution is constructed in the following manner:

Step 1. Randomly draw two samples, of fixed sizes, from two populations with equal means. (These samples don't have to be equal sizes.)

Step 2. Calculate the mean of each sample and record the mean difference.

Step 3. Return the samples to their respective populations.

Step 4. Repeat steps 1-3 until all possible pairs of samples have been drawn.

Step 5. Place the mean differences in a relative frequency distribution.

The result is a random sampling distribution of the difference between means. This distribution of differences has several invariable properties related to the original populations.

The Mean

If we compute the mean of all the differences we collected, we would find it is zero. This makes sense because we drew our samples from two populations whose mean difference is zero.

Staying with our usual notation:

$\overline{X}_1 - \overline{X}_2$ is the difference between sample means.

$\mu_{\overline{x}_1} - \mu_{\overline{x}_2}$ is the mean of the sampling distribution of the difference.

$\mu_1 - \mu_2$ is the mean of the population of differences.

The means of the raw score populations are equal, so the mean of the distribution of differences is also zero.

$$\mu_{\overline{x}_1} - \mu_{\overline{x}_2} = 0$$

$$\mu_1 - \mu_2 = 0$$

The Variance

Because we have a frequency distribution of values (i.e., differences), we can calculate the variance of that distribution. How much do the differences vary around the mean difference of zero? The variance of the sampling distribution of the difference between means is related to the variances of the populations. It is smaller than in the original populations because we are dealing with sample means, not raw scores. Like the sampling distribution of the mean, as the samples become larger, the variance of the sampling distribution of the difference becomes smaller. The variance of the sampling distribution of the difference is related to the population variances in the following manner:

$$\sigma^2_{\bar{X}_1 - \bar{X}_2} = \sigma^2_{\bar{X}_1} + \sigma^2_{\bar{X}_2} = \frac{\sigma_1^2}{n_1} + \frac{\sigma_2^2}{n_2}$$

Variance of the Random Sampling Distribution of the Difference

When we compute the variance of the sampling distribution of the difference from the sums of squares of the groups, the computational formula is

$$\sigma^2_{\bar{X}_1 - \bar{X}_2} = \frac{\Sigma x_1^2 + \Sigma x_2^2}{n_1 + n_2}\left(\frac{1}{n_1} + \frac{1}{n^2}\right)$$

$$\text{or} = \frac{\Sigma x_1^2 + \Sigma x_2^2}{n(n)} \quad \text{if } n_1 = n_2$$

Computational Formulas for the Variance of the Sampling Distribution of the Difference

If the populations from which the samples were drawn have equal variances as well as equal means, and if the samples were of equal sizes, then the relationship between the population variance and the variance of the sampling distribution is $\sigma^2_{\bar{X}_1 - \bar{X}_2} = 2\sigma^2/n$

To summarize:

σ_1^2 is the variance of the first population.
σ_1^2 is the variance of the second population.

$\sigma^2_{\bar{x}_1 - \bar{x}_2}$ is the variance of the sampling distribution of differences between means. The relationship between them is

$$\sigma^2_{\bar{x}_1 - \bar{x}_2} = \frac{\sigma_1^2}{n_1} + \frac{\sigma_2^2}{n_2}$$

The Standard Deviation

You will recall that the standard deviation of a random sampling distribution is called a standard error. The standard deviation of the sampling distribution of the difference is called the **standard error of the difference between means.** The standard error is simply the square root of the variance. To summarize:

σ_1 is the standard deviation of the first population.
σ_2 is the standard deviation of the second population.

$\sigma_{\bar{x}_1 - \bar{x}_2}$ is the standard deviation of the sampling distribution of differences between means: The standard error is related to the populations in the following way:

$$\sigma_{\bar{x}_1 - \bar{x}_2} = \sqrt{\sigma^2_{\bar{x}_1} + \sigma^2_{\bar{x}_2}} = \sqrt{\frac{\sigma_1^2}{n_1} + \frac{\sigma_2^2}{n_2}}$$

Standard Error of the Random Sampling Distribution of the Difference

$$= \sqrt{\frac{2\sigma^2}{n}}$$

if $\sigma_1 = \sigma_2$ and $n_1 = n_2$

The computational formulas for the standard error of the difference are

$$\sigma_{\bar{x}_1 - \bar{x}_2} = \sqrt{\frac{\Sigma x_1^2 + \Sigma x_2^2}{n_1 + n_2}\left(\frac{1}{n_1} + \frac{1}{n^2}\right)}$$

$$\text{or} = \sqrt{\frac{\Sigma x_1^2 + \Sigma x_2^2}{n(n)}} \quad \text{if } n_1 = n_2$$

Computational Formulas for the Standard Error of the Difference

The Shape

Remember the Central Limit Theorem? It governs the shape of the sampling distribution of the difference in the same way it did with the sampling distribution of the mean. This distribution will be normal if the populations are normal or if the samples sizes are reasonably large. We can use the normal curve to solve problems about the sampling distribution of the difference as we did with the sampling distribution of the mean.

Solving Problems with the Random Sampling Distribution of the Difference

Once again, to use the normal curve table, we need to convert our statistic to its Z-score equivalent. The mean of the sampling distribution is subtracted from its corresponding sample statistic, and this difference divided by the standard deviation of the sampling distribution of the statistic. Since our statistic is the mean difference, the Z-score formula is

$$Z = \frac{(\overline{X}_1 - \overline{X}_2) - (\mu_{\overline{X}_1} - \mu_{\overline{X}_2})}{\sigma_{\overline{X}_1 - \overline{X}_2}} \qquad \text{Z for the difference}$$

Often the right part of the numerator is omitted since the mean of the sampling distribution of the difference is equal to the mean difference in the population: zero.

We can solve problems about mean differences in the same way we solved problems with single means when the shape of the distribution of mean differences is normal. Consider two raw score populations with means of 100 and standard deviations of 10. Let us randomly select a sample of 25 observations from each population and compute the difference between the sample means.

How often would we expect to get mean differences between 5 and −5? Using the same procedure, we take these steps:

Step 1. Draw the curve.

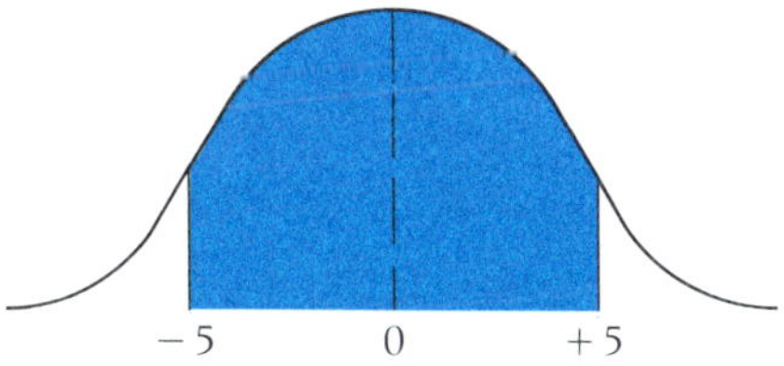

Step 2. Compute the Z scores.

$$Z = \frac{\overline{X}_1 - \overline{X}_2}{\sigma_{\overline{X}_1 - \overline{X}_2}}$$

Since we have drawn equal-sized samples from populations with equal variances, we calculate the standard error by:

$$Z \text{ of } \pm 5 = \frac{\pm 5}{\sqrt{\frac{2(10)^2}{25}}} = \frac{\pm 5}{2.83} = \pm 1.77$$

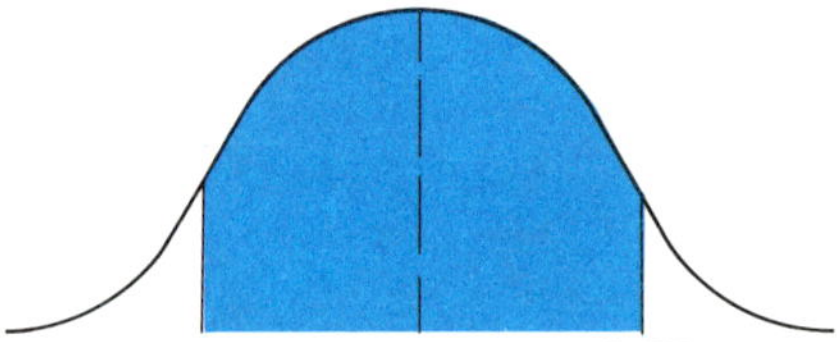

These two Z-scores correspond to mean differences of 5 points or less between the two samples in either direction.

Step 3. Look up the area in the normal curve tables.

Area between the mean of 0 and a Z of ± 1.77 is $2(0.4616) = 0.9232$

About 92% of the time we would expect to get samples whose means differ by 5 points or less either way, when drawing from such populations. This much sampling fluctuation, then, is quite likely.

Now let's try another example? Beyond what two values would we expect only the extreme 5% of the mean differences to fall, $2\,^1/_2$% on either side?

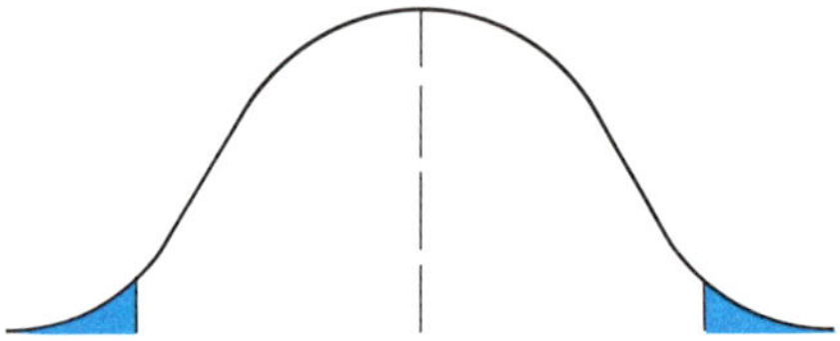

Step 2. Look up the Z values which separate the middle 95% from the extreme 5% of the area.

$$Z = \pm 1.96$$

Step 3. Convert the Z-scores to mean differences by rearranging the formula.

$$\overline{X}_1 - \overline{X}_2 = (\mu_{\overline{X}_1} - \mu_{\overline{X}_2}) \pm Z(\sigma_{\overline{X}_1 - \overline{X}_2})$$
$$= 0 \pm 1.96(2.83) = \pm 5.55$$

We would expect our sample means to differ 5.55 points or more in either direction, 5% of the time. Solving problems about mean differences follows the same logic as solving problems about single means. The Central Limit Theorem assures us that the shape of the sampling distribution of the difference will be normal if the populations are normal, or close to normal as long as the samples are reasonably large.

The Random Sampling Distribution of a Proportion

Recall our coin toss experiment. In one toss of a coin, only two outcomes are possible. You will recall from Chapter 7 that this is called a *dichotomous variable*. Dichotomous variables are ones where only two mutually exclusive outcomes are of interest. Up to now we have been discussing random sampling distributions where means of samples have been computed. Dichotomous variables can be placed in a random sampling distribution as well. We'll call this **the random sampling distribution of a proportion**. It's easier to say proportion, and any dichotomous variable can be thought of in terms of proportions.

We would construct the random sampling distribution of a proportion in the following manner:

Step 1. Randomly select a sample of some fixed size from the population of a dichotomous variable.

Step 2. Determine the proportion of times the value of interest occurs.

Step 3. Return the sample to the population.

Step 4. Repeat steps 1–3 until all possible samples have been drawn.

Step 5. Place these sample proportions in a relative frequency distribution.

If we were interested in setting up the random sampling distribution of our simple but illustrative experiment of 6 tosses of a coin, we would:

Step 1. Toss the coin 6 times. (Our sample size is 6.)

Step 2. Count the number of times a head occurs and record this as a proportion (e.g., 0.5 or 3/6)

Step 3. Repeat the first two steps an infinite number of times.

Step 4. Place our proportions in a relative frequency distribution.

If you suspect that this distribution has certain invariable properties, you are quite correct.

The Mean

In our coin toss experiment the outcome that will occur most frequently is 3 heads in 6 tosses. With a fair coin, we would expect to get half heads and half tails more often than any other outcome. Occasionally, we will get an unlikely outcome, such as 6 heads in a row, but not often. The mean of the random sampling distribution of a proportion (**P_p**) is equal to the proportion of times the outcome of interest is expected to occur in the population (**P**). In the case of a coin, we would expect to get heads 50% or 0.50 of the time. This is also the probability of a head; proportions can be thought of as probabilities. (You may recall our discussion of a relative frequency distribution as a probability distribution.) We will use lower case **p** to refer to the proportion of times the outcome of interest occurred in the sample. The mean of all the sample proportions determined from all possible samples is equal to the expected value of the outcome of interest in the population (P). To summarize:

P is the proportion for the outcome of interest in the population or the probability of the outcome of interest.

p is the proportion of times the outcome of interest occurred in the sample.

P_p is the mean of the random sampling distribution of the proportion.

$P_p = P$

The Variance

All the sample proportions vary around the mean of the distribution. The variance of this sampling distribution is

$\sigma_p^2 = PQ/n$ Variance of the Sampling Distribution of a Proportion

If we think of P as the probability of the outcome of interest, then Q is the probability of the other outcome. Q is always equal to $1 - P$.

The Standard Deviation

As you might expect, the standard deviation of the random sampling distribution of a proportion is called the **standard error of the proportion**. It is the square root of the variance.

$$\sigma_p = \sqrt{PQ/n} \qquad \text{Standard Error of the Proportion}$$

The Shape

Once again the Central Limit Theorem governs the shape of the sampling distribution of a proportion. When the sample size is large, the distribution is normal in shape. Let's repeatedly toss our coin 6 times in a row. Now, let's repeatedly toss the coin 50 times in a row. The relative frequency distribution constructed in the second experiment would be closer to the normal distribution than the one constructed in the first experiment. A rule of thumb is that samples must be about 40 or larger for the sampling distribution to be normal enough to use the normal curve as a model.

Solving Problems with the Random Sampling Distribution of a Proportion

Let's toss our coin 50 times and record the number of heads we get. If we did this over and over again, how often would we expect to get 40 or more heads in 50 tosses? We can solve this problem using the same procedure we used earlier. To use the normal curve table to find areas, we must have a Z-score. Since the general formula for Z is to subtract the mean of the sampling distribution from the sample statistic and divide this difference by the standard error, our Z formula is

$$Z = \frac{p - P_p}{\sigma_p} \qquad \text{Z for a Proportion}$$

Now let's apply the procedure.

Step 1.

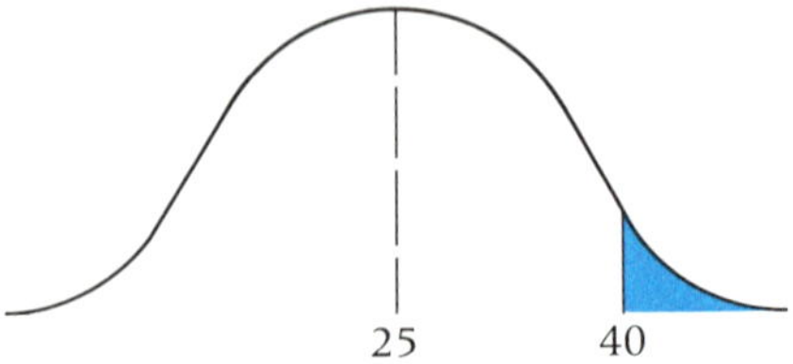

Since we are tossing our coin 50 times, 50 is our sample size. The mean of the sampling distribution of the proportion (P_p) equals the expected proportion in the population. If the coin is fair we would expect to get heads on half of the tosses, or 25 times in 50 tosses, on the average.

Step 2. Compute the Z-score. Our sample proportion is 40/50 or 0.80. The mean of the sampling distribution is 25/50 or 0.50.

$$Z = \frac{p - P_p}{\sigma_p} = \frac{0.80 - 0.50}{\sqrt{(0.50)(0.50)/50}} = \frac{0.30}{0.07} = 4.29$$

Step 3. Look up the Z value. Once again our Z value is too large for the table. It is extremely unlikely to get 40 heads in 50 tosses of a fair coin. This outcome would occur so rarely that our best bet is that the coin is not fair.

Let's look at a different kind of problem. Remember Mendel's Law? Consider a variety of garden pea where Mendel's Law states that 25% of these peas would be pink in colour and the rest should be green. If we were to randomly select many samples of 400 peas each, how often would we expect to get 79 or fewer pink peas out of the 400 we picked?

Step 1.

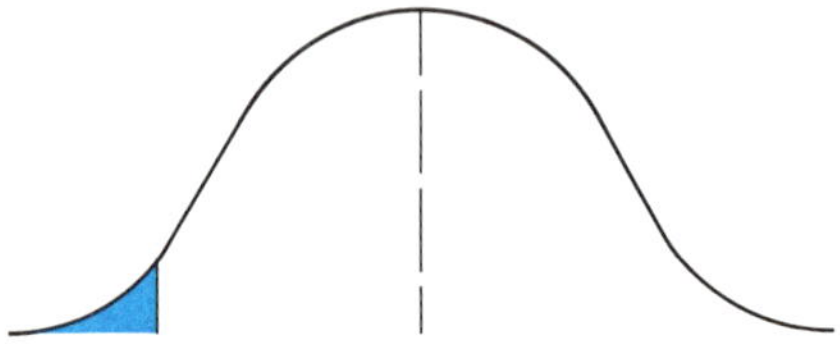

The mean of the sampling distribution is 0.25 because Mendel's Law states that 25% or 100 of the 400 peas should be pink. We are interested in the area under the curve for proportions of 79/400 or 0.198 or less.

Step 2.

$$Z = \frac{0.198 - 0.25}{\sqrt{(0.25)(0.75)/400}} = -\frac{0.052}{0.022} = -2.36$$

Step 3.

$$\text{Area} = 0.0091$$

We would expect to get 79 or fewer pink peas out of 400 less than 1% of the time.

Knowing the shape of the random sampling distribution of a statistic allows us to determine the likelihood of different kinds of sample outcomes. Here we have looked at the normal distribution. Later, we will discuss sampling distributions of other statistics which are not normally shaped. The procedure will not change, however. When we know the frequency of the expected outcomes, then we can make informed decisions about particular outcomes.

The normal curve is a very useful tool in inferential statistics. It can be used to approximate several different random sampling distributions.

SUMMARY OF TERMS AND FORMULAS

Random Sampling Distribution of:	Mean	Variance	Stand. Dev.	Z formula
The Mean	$\mu_{\bar{x}}$	$\sigma^2_{\bar{x}}$	$\sigma_{\bar{x}}$	$\frac{\bar{X} - \mu_{\bar{x}}}{\sigma_{\bar{x}}}$
The Difference	$\mu_{\bar{x}_1} - \mu_{\bar{x}_2}$	$\sigma^2_{\bar{x}_1 - \bar{x}_2}$	$\sigma_{\bar{x}_1 - \bar{x}_2}$	$\frac{\bar{X}_1 - \bar{X}_2}{\sigma_{\bar{x}_1 - \bar{x}_2}}$
The Proportion	P_p	σ^2_p	σ_p	$\frac{p - P_p}{\sigma_p}$

Computational Formulas for the Standard Error of the Difference

$$\sigma_{\bar{x}_1 - \bar{x}_2} = \sqrt{\frac{\Sigma x_1^2 + \Sigma x_2^2}{n_1 + n_2}\left(\frac{1}{n_1} + \frac{1}{n_2}\right)}$$

$$\sigma_{\bar{x}_1 - \bar{x}_2} = \sqrt{\frac{\Sigma x_1^2 + \Sigma x_2^2}{n(n)}} \quad \text{if } n_1 = n_2$$

Relationship between Properties of the Random Sampling Distribution and Properties of the Population.

Random Sampling Distribution

R.S.D. of the Mean $\mu_{\bar{x}} = \mu$

$\sigma_{\bar{x}} = \sigma/\sqrt{n}$

R.S.D. of the Difference $\mu_{\bar{x}_1} - \mu_{\bar{x}_2} = \mu_1 - \mu_2 = 0$

$$\sigma_{\bar{x}_1 - \bar{x}_2} = \sqrt{\frac{\sigma_1^2}{n_1} + \frac{\sigma_2^2}{n_2}}$$

R.S.D. of the Proportion $P_p = P$

$$\sigma_p = \sqrt{PQ/n}$$

EXERCISES

1. The following data are a population of raw scores.

Data: 1, 2, 4, 7, 11

Compute the mean and standard deviation of this population.

2. Using the population of scores in question 1, list the 25 possible samples of size 2 which can be drawn from this population with replacement. Find the mean of each sample.

3. Place the sample means from question 2 into an absolute frequency distribution. Find the mean of the distribution and the standard deviation of the distribution using the following formulas: (Hint: N = 25)

$$\mu_{\bar{x}} = \Sigma f\bar{X}/N$$

$$\sigma_{\bar{x}} = \sqrt{\frac{\Sigma f\bar{X}^2 - (\Sigma f\bar{X})^2/N}{N}}$$

You will find that the mean of the sample means ($\mu_{\bar{x}}$) is the equal to the mean of the original population that you computed in question 1(μ). Check to make sure that the standard deviation or standard error you computed on your distribution of sample means is equal to $\sigma/\sqrt{n}$, where n is the sample size of 2.

4. Using the distribution of sample means from question 3, determine the probability of drawing at random a single sample of 2 scores whose mean is:

a) equal to 5
b) > 5
c) > 4 and < 6

5. A distribution of scores has a mean of 75 and a standard deviation of 15. If all possible samples were drawn with replacement from this population, what will be the values of $\mu_{\bar{x}}$ and $\sigma_{\bar{x}}$ if:

a) n = 225
b) n = 100
c) n = 25

6. What is the probability of drawing from the population in question 5, a single sample of 25 scores whose mean is between 70 and 80?

CHAPTER 9

Inference with the Normal Curve

Chapters 6, 7, and 8 discussed topics which are basic to inferential statistics. You now know the properties of the normal curve and several random sampling distributions. You understand probability and how it applies to frequency distributions. It's now time to apply this knowledge to make inferences about populations from samples drawn from them.

Hypothesis Testing and Interval Estimation

> **a priori:** Before the fact.
> **post hoc:** After the fact.

Hypothesis testing and **interval estimation** are two inferential techniques which are based on the concept of random sampling distributions. They both use sample statistics to make inferences about population parameters.

In hypothesis testing, we have an **a priori** (before the fact) hypothesis about the value of some parameter or the relationship between parameters. Sample data are used to determine whether or not this hypothesis is reasonable. The sample data are used to "test our guess."

With interval estimation, we do not have any a priori hypotheses about the value of the parameter(s). We use the sample data to infer what the value of the parameter might be. Interval estimation, then, is a **post hoc** (after the fact) technique. Because interval estimation is a simpler technique, it will be discussed first.

Interval Estimation with the Normal Curve

When we use interval estimation with the normal curve, we are asking "What is the value of this population parameter?" or "What is the relationship between parameters?" Interval estimation provides us with a method to set up a *range of values* within which the parameter will likely fall. Although the size of the range of values is arbitrary, most researchers in psychology and related disciplines use two ranges, the 95% and the 99% **confidence intervals**.

Setting up the Confidence Interval for a Population Mean

Critical value: The value of a statistic corresponding to a given significance level determined by its sampling distribution.

In Chapter 8 you saw that a distribution of means, calculated from random samples drawn from a population, distributes itself normally within the limits of the Central Limit Theorem. A single sample mean, if selected randomly, may be viewed as a *single random case* drawn from the distribution of all sample means. Recall that in the random sampling distribution of the mean, 95% of the sample means fall within ± 1.96 standard deviations of the mean of the sampling distribution, and 99% of the means fall within ± 2.58 standard deviations of the mean of this distribution. These two values, ± 1.96 and ± 2.58, are called **critical values** in inferential statistics.

When we know the population standard deviation and the mean of a single random sample, we can estimate where the population mean (μ) is likely to fall. The formulas for setting up the 99% and the 95% confidence intervals (CI) are

Confidence Intervals for the Mean

95% CI	$\bar{X} \pm 1.96\ \sigma_{\bar{X}}$
99% CI	$\bar{X} \pm 2.58\ \sigma_{\bar{X}}$

Let's examine a specific example. Suppose we have drawn a random sample from a population whose standard deviation is 15. We find that the mean of our 25 observations is 100. We wish to determine a range of values within which we are reasonably sure that the population mean will fall. Likely the population mean will not equal 100, since we know that sample means vary somewhat from the population mean. If we claimed that $\mu = 100$, the chances are we would be wrong. To make a better guess, use the two formulas above to provide the 95% and 99% CI's.

Data: $\sigma = 15$ $\quad\quad \overline{X} = 100$
$n = 25$

Before we use the formulas, we need to find the standard deviation of the sampling distribution (the standard error).

$$\sigma_{\overline{X}} = \sigma/\sqrt{n}$$
$$\sigma_{\overline{X}} = 15/\sqrt{25} = 3$$
$$95\%\ \text{CI: } \overline{X} \pm 1.96\sigma_{\overline{X}}$$
$$= 100 \pm 1.96(3) = 100 \pm 5.88 = 94.12 \text{ and } 105.88$$

We are 95% confident that the population mean lies between 94.12 and 105.88. In other words, our confidence that the population mean is greater than or equal to 94.12, and less than or equal to 105.88, is .95. This can be expressed in notation:

$$C(94.12 \leq \mu \leq 105.88) = .95$$

If we wanted to be more confident, we could set up a 99% confidence interval by:

$$99\%\ \text{CI: } \overline{X} \pm 2.58\sigma_{\overline{X}}$$
$$= 100 \pm 2.58(3) = 100 \pm 7.74 = 92.26 \text{ and } 107.74$$

and express this in notation as

$$C(92.26 \leq \mu \leq 107.74) = .99$$

As you can see, the 99% CI sets a wider range of values within which the population mean is likely to fall.

Why do we use the term "confidence" instead of "probability"? This is a good question. When we set up the 95% CI, for example, we know that 95% of the sample means will lie within 1.96 standard error units from the mean of the sampling distribution. Therefore, 5% of the sample means will be further away from $\mu_{\overline{X}}$ than $1.96\sigma_{\overline{X}}$. For these few sample means, the limits set up by our procedure will *not* encompass μ. (Remember that $\mu_{\overline{X}} = \mu$). In other words, 5% of the time, we will obtain a sample mean so far from the mean of the sampling distribution, that our confidence interval will not include μ. *Before* we draw our sample to set up the 95% CI, we can say that the *probability* that our interval will contain μ is .95. However, once the data have been collected and we have set our limits, a given interval *either does or does not* contain μ. Probability is a term that we use to express the likelihood of future outcomes. Once the event has taken place, we cannot use the term probability. A good analogy is one toss of a fair coin. While the coin is still up in the air, we can say the probability of getting a head is 1/2. Once the coin has landed, however, either a head or a tail has occurred. The event has taken place and we can no longer use the term probability. When setting up limits for μ, then, we use the term *confidence* rather than probability.

With any inferential technique, there is some probability of error. When we set up confidence intervals, there is always a possibility that our limits will not encompass the parameter of interest.

Consider a population of raw scores with a mean of 100 and a standard deviation of 20. If we randomly draw all possible samples of 100 observations from this population, and place the sample means in a relative frequency distribution, we will have the random sampling distribution of the mean for a sample size of 100. Suppose we draw a single sample of 100 observations from the population and compute a sample mean of 103. Let's go ahead and set up the 95% CI for this sample mean.

$$95\%\ \text{CI} \qquad 103 \pm 1.96(20/\sqrt{100})$$

$$= 103 \pm 3.92 = 99.08 \text{ and } 106.93$$

We are 95% confident that the population mean lies between these two values and justifiably so, since we know that the population mean is 100.

Now, let's draw another sample from the population and compute the mean of the sample. We find it is 95. Now we set up the 95% CI for this sample mean.

$$95\%\ \text{CI} \qquad 95 \pm 1.96(2) = 91.08 \text{ and } 98.92.$$

We would claim that μ lies between these two values. Are we correct? Not this time. The limits do not include μ. The following figure illustrates what happened.

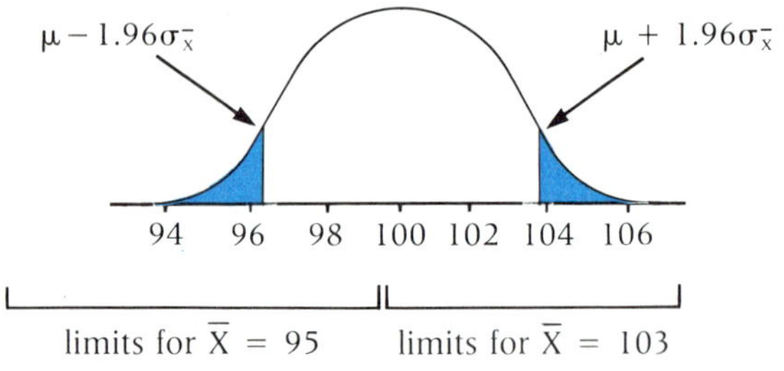

Look at the range of values we set up for the two sample means. You can see that for the mean of 103, our limits included μ, but for the mean of 95, they did not. Whenever a sample mean is further away from the mean of the sampling distribution than $\pm 1.96\, \sigma_{\bar{X}}$, the limits set up by the 95% CI will not contain the population mean. If, on the other hand, the sample mean is within $\pm 1.96\, \sigma_{\bar{X}}$ of $\mu_{\bar{X}}$, the population mean will be included in our interval. We know that 5% of the time a single sample mean will be further away than $\pm 1.96\, \sigma_{\bar{X}}$ of $\mu_{\bar{X}}$, and so 5% of the time we make an incorrect inference about the location of the population mean. Obviously, if we want to be very careful about error, we would use a stricter criterion. The 99% CI is more conservative and we will be in error only 1% of the time.

Setting up the Confidence Interval for the Difference between Population Means

Similarly, confidence intervals can be determined for the difference between population means. The random sampling distribution of the difference tells us what kinds of differences we can expect between sample means when they come from populations with equal means. If we draw two random samples and compute the mean difference, we can set up a range of values between which we are reasonably confident the mean difference in the population lies. The formulas for 95% and 99% confidence intervals for the difference are

Confidence Intervals for the Difference	
95% CI	$(\bar{X}_1 - \bar{X}_2) \pm 1.96(\sigma_{\bar{X}_1 - \bar{X}_2})$
99% CI	$(\bar{X}_1 - \bar{X}_2) \pm 2.58(\sigma_{\bar{X}_1 - \bar{X}_2})$

Again, we are simply using the sample statistic, the mean difference, to set up a range of values within which we think the population parameter $(\mu_1 - \mu_2)$ lies. Ready for an example?

In an Introductory Psychology class, 16 students are randomly selected and assigned to two experimental conditions. Both groups are given the same 25-item multiple-choice test. Group 1 is given hints on taking multiple-choice tests; Group 2 is not. We want to set up the 99% CI for the difference. Here are the data.

$$\text{DATA: } \overline{X}_1 = 13.25 \qquad \overline{X}_2 = 14.75$$
$$\Sigma x_1^2 = 43.5 \qquad \Sigma x_2^2 = 75.5$$
$$n_1 = n_2 = 8$$
$$99\%\ \text{CI} \qquad (\overline{X}_1 - \overline{X}_2) \pm 2.58(\sigma_{\overline{x}_1 - \overline{x}_2})$$

We first need to compute the standard error of the difference. Since the samples are the same size, we can use:

$$\sigma_{\overline{x}_1 - \overline{x}_2} = \sqrt{\frac{\Sigma x_1^2 + \Sigma x_2^2}{n(n)}} = \sqrt{\frac{43.5 + 75.5}{8(8)}} = 1.36$$

Now we can go ahead and set up the confidence interval.

$$99\%\ \text{CI}\ (13.25 - 14.75) \pm 2.58(1.36) = -1.50 \pm 3.51$$
$$= 2.01 \text{ and } -5.01$$

We are 99% confident that the true difference in the population lies somewhere between −5.01 and 2.01 or, expressed in notation:

$$C(-5.01 \leq \mu \leq 2.01) = .99$$

Setting up the Confidence Interval for the Population Proportion

We can set up confidence intervals for the proportion using the same logic and procedure. The formulas are

> Confidence Intervals for the Proportion
>
> 95% CI $\quad p \pm 1.96(\sigma_p)$
>
> 99% CI $\quad p \pm 2.58(\sigma_p)$

In this case, we are setting up a range of values within which we are reasonably sure that the population proportion lies. Let's do an example.

You have two cards, one red and one black. You randomly select 52 friends and ask each to guess the colour of the card you are holding behind your back. Thirty-eight people guess correctly. You are interested in knowing the proportion of people in the population who would guess correctly. You decide to set up the 95% CI. For each subject, we assume that the probability of guessing correctly is 1/2.

DATA: $p = 38/52 = 0.73$

$$\sigma_p = \sqrt{PQ/n} = \sqrt{(0.5)(0.5)/52} = 0.07$$

$$95\%\ \text{C.I.} = 0.73 \pm .1.96(0.07) = 0.59 \text{ and } 0.83$$

You are 95% confident that the true percentage in the population is between 59% and 83%, at least under the conditions of your experiment. We might wonder about how you conducted this study, however!

Setting up confidence intervals can be done for many population parameters. All we need is the sample statistic, the standard deviation of the sampling distribution of that statistic, and knowledge of the shape of that sampling distribution. The general formula is

CI for parameter = statistic ± critical value (standard error)

We use confidence intervals when we have no a priori hypothesis about the value of the population parameter. When we have an hypothesis about some parameter and we wish to test its validity we use a different inferential technique.

Hypothesis Testing with the Normal Curve

In research, we often have hypotheses about events or phenomena. These "educated guesses" may come from theories or from previous research. This section discusses the statistical approach for "testing our guess."

Types of Hypotheses

Three levels of hypotheses can be described in terms of the degree of quantification involved.

Conceptual Hypotheses

A **conceptual hypothesis** is a statement about the relationship between theoretical concepts. Hypotheses like, "Punishment facilitates learning", or "Intelligence is negatively related to sex appeal" are examples of conceptual

Operationalize: To make observable or measurable.

hypotheses. Punishment, learning, intelligence, and sex appeal are theoretical concepts. They can never be directly tested because they cannot be measured. Conceptual hypotheses must be **operationalized** or made measurable before we can test them.

Research Hypotheses

Experimental hypothesis: States relationship between an independent variable and a dependent variable.

Correlational hypothesis: States relationship between dependent variables.

A **research hypothesis** is a statement about the expected relationship between *observable* or *measurable* events. An **experimental** research hypothesis states expected relationships between independent and dependent variables. For example, "Shock, following errors, will decrease the numbers of errors made" is an experimental research hypothesis. A **correlational** research hypothesis states the expected relationship between two or more dependent variables. "The higher the score on a standard IQ test, the lower the rating of sex appeal by 10 male judges" is an example of a correlational research hypothesis. A research hypothesis restates the conceptual hypotheses in observable or measurable terms.

Statistical Hypotheses

A **statistical hypothesis** states an expected relationship between numbers representing statistical properties of data (e.g., mean, variance, correlation). This type of hypothesis is always a guess about the value of a population parameter or about the relationship between values of two or more parameters, at least in parametric hypothesis testing.

Examples of statistical hypotheses include, "The mean number of errors is the same under shock and no-shock conditions" and "The correlation between IQ and sex appeal rating is zero."

Researchers usually begin with a conceptual hypothesis derived from some theory of published work. They then determine the best way to collect observations or measurements appropriate to their conceptual hypothesis. Data collection follows and statistical procedures commence.

The Logic and Procedure for Testing a Conceptual Hypothesis

The steps for testing hypotheses about theoretical concepts are as follows:

Step 1. Restate the conceptual hypothesis as a research hypothesis. (Operationalize the concepts.)

Step 2. Make a statement about expected values of parameters of interest. (State the statistical hypothesis.)

Step 3. Collect and summarize the data.

Step 4. Test the statistical hypothesis.

Step 5. Make conclusions about the conceptual hypothesis.

We will start at Step 2. A statistical hypothesis includes the hypothesis you wish to disprove (the **null**) and the hypothesis you wish to confirm (the **alternate**).

Null and Alternate Hypotheses

> **Null hypothesis:** States expected value of a parameter or expected relationship between parameters.
>
> **Alternate hypothesis:** States a value or relationship different from the null.

The null hypothesis (**H_o**) is the one you test and hope to prove wrong, reject or nullify. If the null is rejected, the alternate (**H_a**) is accepted. The aim of hypothesis testing is to show that the null is false and, thereby, accept an alternate hypothesis. The alternate hypothesis corresponds with the research hypothesis. Confirmation of the research hypothesis lies in *rejecting the null.* For example, if we wish to show that a coin is biased, we must prove *it is not fair*. This may seem to be a pretty strange way of doing things, but it will become clearer to you, I hope!

The null hypothesis specifies the expected value of a single population parameter or the expected relationship between two or more parameters. Some examples of null hypotheses are:

$$H_o\colon \mu = 100$$
$$P = 1/2$$
$$\sigma_1^2/\sigma_2^2 = 1$$
$$\mu_1 = \mu_2$$

Non-directional alternate: Hypothesis that negates the null.

Directional alternate: Hypothesis that specifies the direction of the difference from the null.

The alternate hypothesis asserts that the value of relationship specified by the null is not true. You test the null hypothesis and if it is rejected, you accept an alternate hypothesis. An alternate hypothesis that simply negates the null is called a **non-directional alternate**. If it specifies the direction of the difference, it is called a **directional alternate**. Some examples are

H_o	Non-Directional H_a	Directional H_a
$\mu = 100$	$\mu \neq 100$	$\mu > 100$ or $\mu < 100$
$\mu_1 = \mu_2$	$\mu_1 \neq \mu_2$	$\mu_1 > \mu_2$ or $\mu_1 < \mu_2$

The type of alternate hypothesis used depends on the research hypothesis. If the researcher's interest is in finding a difference only in a particular direction, then a directional alternate is appropriate. Otherwise, a non-directional alternate hypothesis is used.

Testing the Null

The null hypothesis is tested in a clearly defined step-wise fashion. The steps are:

Step 1. Define the H_o and the H_a.

Step 2. Select a random sample(s) from the population(s) of interest and calculate the value of the statistic that corresponds to the parameter specified in the null.

Step 3. Determine the probability of getting a sample outcome, by chance alone, that is as far or further from the value hypothesized by the null, as the one you obtained. This is done in reference to the sampling distribution of the statistic specified by the conditions of the null hypothesis.

Step 4. If the probability is low, reject the null and accept the alternate. If the probability is high, do not reject the null.

Let's try to analyze this procedure conceptually. The null hypothesis is the one we want to prove wrong. Suppose we believe that women are more intelligent than men! The null hypothesis might state that women are equal in intelligence to men and the alternate might state that they are more intelligent. (Remember that the alternate corresponds to the researcher's hypothesis.) Let's assume, for the moment, that men and women are equal in intelligence. If this is true, then two random samples, one of men and one of women, should have similar mean intelligence scores. Of course, we wouldn't expect the sample means to be exactly the same because samples do vary. The question is "How much greater does the mean of the female sample have to be in order to conclude that women are more intelligent than men?" To answer this question, we need to find out what kinds of mean differences are likely to occur when we draw two samples from populations with equal means. Then, if our sample difference is much larger than what we would expect by chance alone, we may feel free to conclude that, in fact, the populations do not have equal means.

The null hypothesis, then, provides us with a sampling distribution which *we know a lot about*. We know what kinds of sample outcomes are likely to occur under the conditions assumed by the null. If our sample outcome is very different from what we would expect, we say that the null hypothesis is probably wrong and an alternate hypothesis is true.

Step 4 says that we reject the null if the probability of getting a sample outcome as deviant as the one we got is very low. How do we decide if the probability is low enough?

Decision Criteria

Significance level: The level of probability at which we will reject the null.

In the behavioral sciences and related disciplines, researchers tend to use two probability values when testing hypotheses. These levels, symbolized by α (alpha) are called **significance levels**. The 1% level of significance ($\alpha = .01$) allows us to reject the null hypothesis if our sample outcome was likely to have occurred 1% or less, of the time, if the null were true. The 5% level of significance allows us to reject the null if the probability of our outcome occurring is equal to or less than .05.

When we test hypotheses with the normal curve as our sampling distribution, we do not need to determine the exact probability of our sample outcome. We know that the Z-scores of ± 1.96 separate the middle 95% of the distribution from the extreme 5%. In other words, Z values lying beyond ± 1.96 will occur less than 5% of the time. Once we have calculated our Z value, we only need to compare it to ± 1.96. If our Z value is larger numerically than these **critical Z values**, then we know that the probability of occurrence of our outcome is less than .05 and we can reject the null hypothesis.

> **Region of rejection:** Area beyond the critical value. Outcomes lying in this area lead to rejection of a null.
> **Region of acceptance:** Outcomes in this area lead to non-rejection of the null.

The same is true for a significance level of .01. The critical values for $\alpha = .01$ are ± 2.58, since these two values separate the middle 99% from the extreme 1% of the distribution. In summary:

$$\text{Critical Z values: } \alpha = .05 \quad Z_{crit} = \pm 1.96$$
$$\alpha = .01 \quad Z_{crit} = \pm 2.58$$

Critical values, then, cut off the appropriate areas in the tails of the distribution corresponding to the level of significance selected by the researcher. Any obtained Z value which falls beyond these critical values is said to lie within the **region of rejection** or the **critical region**. Z values falling between the critical values are said to lie in the "**region of acceptance**." This is illustrated in the following diagram.

We reject the null when the obtained Z value lies in the critical region. Otherwise, we *fail to reject the null hypothesis*. We never say accept the null, rather we always state that we have failed to prove it wrong. The term "region of acceptance" implies that we do accept the null. It might be more appropriate to term this area the "region of non-rejection."

One and Two-Tailed Tests

Recall that the alternate hypothesis may be either a non-directional alternate and therefore negate the null or it may be a directional alternate and specify the direction of the difference.

The critical region of a non-directional alternate lies in both tails, and the test of the null is a **two-tailed test of significance**. The critical region of a directional alternate is on one side only and we run a **one-tailed test of significance**. We are interested in one side of the distribution only, because we have specified this in the alternate hypothesis. We may reject the null only if our outcome is in the critical region of that tail.

The critical values given earlier were for a two-tailed test with a non-directional alternate hypothesis. Recall that ± 1.96 cuts off 95% of the distribution from the extreme 5% with $2\,^1/_2$% in each tail. These are the critical values for a two-tailed test at $\alpha = .05$. However, if the alternate hypothesis is directional, these critical values won't work.

The significance level of .05 for a one-tailed test requires that the extreme 5% lies in *one tail only*, whichever tail is specified in the alternate. When $\alpha = .01$, we need one Z value which cuts off 99% from the extreme 1% in one tail only. These critical values must be smaller than the corresponding ones for a two-tailed test because all the area is in one tail. Looking in Table 1, we find that the Z values are + or -1.64 ($\alpha = .05$) and + or -2.33 ($\alpha = .01$).

The following diagram illustrates the difference between the critical region for a two-tailed and a one-tailed test of significance.

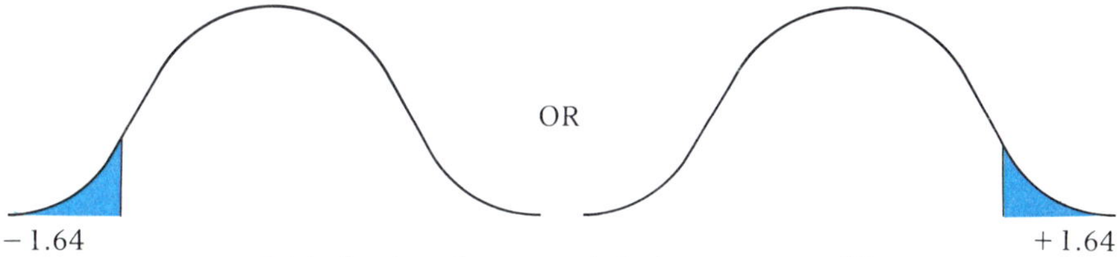

Critical values for a one-tailed test at $\alpha = .05$

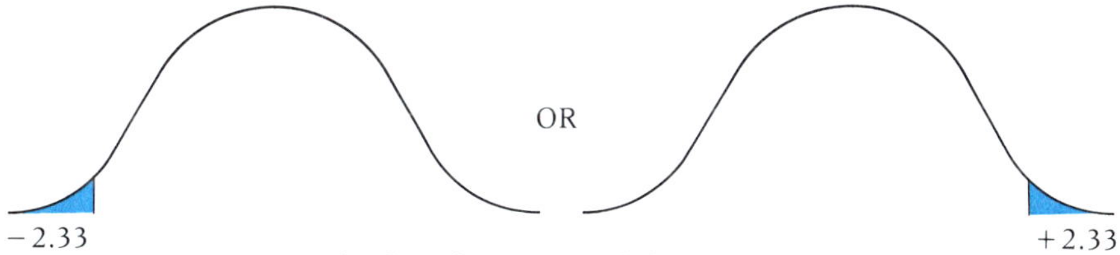

Critical values for a one-tailed test at $\alpha = .01$

In summary:

Critical Z values	$\alpha = .05$	$\alpha = .01$
Two-tailed test	± 1.96	± 2.58
One-tailed test	+ or − 1.64	+ or − 2.33

Notice that the critical values are smaller for a one-tailed test of significance. This means that a smaller Z value is required to reject the null hypothesis. So why don't researchers always do one-tailed tests if it's easier to reject the null?

Part of the answer to this question is that *if our outcome was extremely deviant, but in the wrong direction, we may not reject the null hypothesis!* For example, had we hypothesized that women are smarter than men, and had we found that our female sample mean of IQ scores was very much lower than the mean for the male sample, we would not be able to say that "women are less smart than men." Our outcome was not in the region of rejection specified by the alternate hypothesis. So, although we suspect something is happening here, we may not make the obvious inference, which may be a good thing anyway!

Generally, in research, we use non-directional alternates unless we have absolutely no interest in finding a difference in the other direction, or if finding a difference in the opposite direction to that specified by the alternate *has no real meaning*. For example, if we were interested in testing Clare Voyent's claim that she has ESP, we may run a one-tailed test because a finding that Clare performed significantly *lower* than chance level has no meaning. All we could say was that Clare was a pretty poor guesser—poorer than chance level.

Now that we have covered some of the basics, let's put these skills to work.

Testing an Hypothesis about a Population Mean

Mr. Pierre Theroux wants to compare elementary students in Ontario with those in the rest of the country on a test of mathematical achievement. Specifically, he wonders if his special new math program has had an effect on the students' mathematical skills. The national norms for a standardized math test are $\mu = 85$ and $\sigma = 20$. A randomly selected sample of 100 elementary students from Ontario schools is tested and the mean of the sample is 89. Let's use our steps to help Pierre answer his question. We will test the null at $\alpha = .05$.

Step 1. State the null and the alternate. Remember that the null must state that the students are *not* different from the national norm. The alternate states that they are different. The mean of the population on the math test is 85 and so:

$$H_o: \mu = 85$$
$$H_a: \mu \neq 85$$

Step 2. Draw a random sample from the population and calculate the statistic corresponding to the null hypothesis. We have done this and the sample statistic is:

$$\overline{X} = 89$$

Statistical significance: An outcome leading to rejection of the null is said to be a statistically significant outcome.

Step 3. Determine the probability of getting an outcome as far or farther from the hypothesized value as the one obtained. We don't actually have to do this, rather, we simply compare our value to the critical value. To do this, we need to convert our sample mean to Z-score so we can compare it to the critical Z value. Recall the Z formula for a sample mean:

$$Z = \frac{\overline{X} - \mu_{\overline{X}}}{\sigma_{\overline{X}}}$$

$$Z = \frac{89 - 85}{20/\sqrt{100}} = 2.00$$

Step 4. If the probability is low enough, reject the null hypothesis. This is the decision step. We can see that our obtained Z value is larger than the critical value, 1.96, so we reject the null hypothesis at $\alpha = .05$. Our sample outcome is **statistically significant**; we have evidence that the Ontario students are *significantly different* from the national norm on the test of mathematical achievement.

Summary of Z-Test for a Single Mean

Hypotheses

$\mathbf{H_o}$: μ = specified value
$\mathbf{H_a}$: $\mu \neq <, >$ specified value

Assumptions

1. Subjects are randomly selected.
2. Population distribution is normal.
3. Population standard deviation is known.

Decision Rules

If $Z_{obt} \geq Z_{crit}$, reject the H_o
If $Z_{obt} < Z_{crit}$, do not reject H_o

Formula

$$Z = \frac{\overline{X} - \mu}{\sigma/\sqrt{n}}$$

What we have really found is that the students in the Ontario sample should not have produced a mean so deviant from the national mean, if they were the same as the students in the rest of the country. We conclude that they are not the same. To prove that a sample is from a different population, we must prove that it does not behave in the way a sample from the hypothesized population would, and so it is unlikely to be from the hypothesized population. In this example, we had one sample and we made an inference about the mean of the population from which the sample was drawn. Often, in research, we have two samples and our interest is in the difference between the means of their respective populations.

Testing an Hypothesis about the Difference Between Population Means

To test whether two samples come from populations with different means, we assume a null hypothesis that the samples come from populations with equal means. Even if we drew two random samples from the same population, we would not expect to get identical sample means because samples vary. The question is "How different would we expect sample means to be when they are drawn from populations with equal means?" Once we know what kinds of mean differences to expect, under these conditions, we can compare a specific obtained difference with the expected difference. If our sample means differ more than chance alone would predict, we may conclude that our samples come from populations with unequal means. We would reject the null hypothesis and claim a significant difference between the population means from which our samples were drawn.

To find out what kinds of sample mean differences would occur, with repeated sampling from populations with equal means, we need the random sampling distribution of the difference between means so that we can compare our obtained difference with that distribution. Recall that this distribution is normal within the limits of the Central Limit Theorem, and so the normal curve is an appropriate model for this test.

If the null hypothesis is true and the population means don't differ, we would expect the mean difference of our two samples to be close to zero. If it is far from zero we would claim the null is false and accept the alternate hypothesis. Hypothesis testing for the difference follows the same general procedure that we have been using. We must compute a Z-score for the difference and compare our obtained Z value with the critical values. The critical values for the test for the difference are the same as those we used for the test for a single mean.

Recall the Z formula for the difference between means:

$$Z = \frac{\overline{X}_1 - \overline{X}_2 - (\mu_1 - \mu_2)}{\sigma_{\overline{X}_1 - \overline{X}_2}}$$

and the formulas for the standard error of the difference:

$$\sigma_{\bar{X}_1-\bar{X}_2} = \sqrt{\sigma^2_{\bar{X}_1} + \sigma^2_{\bar{X}_2}} \qquad \text{or}$$

$$\sigma_{\bar{X}_1-\bar{X}_2} = \sqrt{\frac{\sigma_1^2}{n_1} + \frac{\sigma_2^2}{n_2}} \qquad \text{or}$$

$$\sigma_{\bar{X}_1-\bar{X}_2} = \sqrt{\frac{2\sigma^2}{n}} \qquad \text{if } \sigma_2^2 = \sigma_2^2 \text{ and } n_1 = n_2$$

Ready for an example? A school board member is interested in the outbreak of "free" high schools where students are given a much more flexible curriculum and allowed to study independently. She decides to select a random sample of 25 students from a "free" school and another sample of 25 students from a traditional high school. All students are given a standardized achievement test which has a standard deviation of 15. The mean score of the "free" school students on this test is 145 and the mean score for the traditional high school students is 148. Let's test the hypothesis that "free" school and traditional school students don't differ in achievement. We will use $\alpha = .01$.

Step 1. State the null and the alternate.

H_o: $\mu_1 = \mu_2$ — This is always the null for a test of the difference.

H_a: $\mu_1 \neq \mu_2$ — We will use a non-directional alternate.

Step 2. Draw two random samples from the populations and calculate the statistic corresponding to the null hypothesis. We have done this and the sample statistic is:

$$\bar{X}_1 - \bar{X}_2 = 145 - 148 = -3$$

Step 3. Convert the sample statistic to a Z-score and compare it to the critical values. Z_{crit} for a two-tailed test at $\alpha = .01$ is ± 2.58.

$$\sigma_{\bar{X}_1-\bar{X}_2} = \sqrt{\frac{2\sigma^2}{n}} = \sqrt{2(15)^2/25} = 4.24$$

$$Z = \frac{\bar{X}_1 - \bar{X}_2 - (\mu_1 - \mu_2)}{\sigma_{\bar{X}_1-\bar{X}_2}} = \frac{145 - 148 - 0}{4.24} = -0.71$$

Step 4. Make a decision. Our obtained Z value is not larger numerically than the critical value. Because it does not fall in the critical region, we do not reject the null hypothesis. The probability of getting a sample outcome as far or farther from the hypothesized population difference than the one we got is greater than .01. We have no statistical evidence that our samples come from populations with different means. There is no significant difference between "free" school students and students from traditional schools.

Summary of Z-Test for Independent Means

Hypotheses

H_o: $\mu_1 = \mu_2$
H_a: $\mu_1 \neq \mu_2$ or $\mu_1 < \mu_2$ or $\mu_1 > \mu_2$

Assumptions

1. Subjects are randomly selected and independently assigned to groups.
2. Population distributions are normal.
3. Population standard deviations are known.

Decision Rules

If $Z_{obt} \geq Z_{crit}$, reject the H_o
If $Z_{obt} < Z_{crit}$, do not reject H_o

Formula

$$Z = \frac{\overline{X}_1 - \overline{X}_2}{\sqrt{\dfrac{\Sigma x_1^2 + \Sigma x_2^2}{n_1 + n_2}\left(\dfrac{1}{n_1} + \dfrac{1}{n_2}\right)}}$$

Testing an Hypothesis about a Population Proportion

To test an hypothesis about a population proportion we assume, in the null hypothesis, that the population proportion has some specified value and use the same general approach as we used earlier. Let's use our steps to test an hypothesis from Mendelian Law (remember Mendel's peas?). Three hundred garden peas were randomly selected. According to Mendelian Law, 15% of the stock the peas were taken from should exhibit the recessive characteristic of sterility. Of the 300 peas selected, 65 were found to be sterile. Let's test the hypothesis that the peas came from a new hybrid stock. We will use $\alpha = .05$.

Step 1. H_o: $P = 0.15$
H_a: $P \neq 0.15$

Step 2. $p = 65/300 = 0.22$

Step 3.

$$Z = \frac{p - P}{\sqrt{PQ/n}}$$

$$Z = \frac{0.22 - 0.15}{\sqrt{(0.15)(0.85)/300}} = \frac{0.07}{0.02} = 3.5$$

Step 4. Our obtained Z value falls in the critical region so we reject the null hypothesis. We have evidence that the stock from which our sample was selected is significantly different from the stock hypothesized in the null.

Testing the null hypothesis involves determining the probability of a particular sample statistic occurring with respect to the sampling distribution of that statistic. The general approach is to calculate the Z value corresponding to the statistic and compare it to the critical values. The general Z formula we use is

$$Z = \frac{\text{statistic} - \text{hypothesized parameter value}}{\text{standard error}}$$

In summary:

If $Z_{obtained} \geq Z_{critical}$ then $p < \alpha$ and we reject the null.
If $Z_{obtained} < Z_{critical}$ then $p > \alpha$ and we fail to reject the null.

Summary of Z-Test for a Proportion

Hypotheses

H_o: $P =$ specified value
H_a: $P \neq, <, >$ specified value

Assumptions

1. Subjects are randomly selected.
2. Sampling distribution of the statistic is normal.
3. Observations are dichotomous.

Decision Rules

If $Z_{obt} \geq Z_{crit}$, reject the H_o
If $Z_{obt} < Z_{crit}$, do not reject H_o

Formula

$$Z = \frac{p - P}{\sqrt{PQ/n}}$$

Consequences of Statistical Decisions

Whenever we make an inference, there is some probability that we will be wrong! A careful researcher is well aware of the consequences of statistical decisions.

Statistical Significance

The null hypothesis states that a given parameter has some value. When the null is rejected, we claim that the parameter is unlikely to have that value. In other words, there is a *difference between the true population value and the hypothesized population value.*

In practice, any sample outcome that occurs by chance alone 5% (in the case of $\alpha = .05$) or 1% (in the case of $\alpha = .01$) of the time when the null is true, will lead us to conclude that the null is false!

Type I and Type II Errors

A decision as to whether the null is true or false is *never* made with certainty. A **Type I error** occurs when the null hypothesis is true, but our sample outcome leads us to reject it. In other words, we obtained a rare sample outcome that leads us to make an error. Romeo made a Type I error when he wrongly rejected his hypothesis: "Juliet is alive." His error was tragic. The probability of making a Type I error (i.e., the Type I error rate) is set by α. For example, if the null is true and our α level is .05, we will make a Type I error 5% of the time. If α is .01, we will make a Type I error only 1% of the time. Obviously, to reduce the probability of a Type I error, we need to lower the value of α (from .05 to .01, for example).

Unfortunately, it doesn't stop here! A **Type II error** is made when the null is false but we fail to reject it. In other words, a difference between the true parameter value and the hypothesized value exists, but our test did not discover this. Julius Caesar made a Type II error when he wrongly accepted the null hypothesis: "Brutus is my buddy." Hence his response, "Et tu, Brute?" His error, too, was tragic. The probability of a Type II error (i.e., the Type II error rate) is related to alpha and is called **beta** (β). The exact relationship between α and β is not determined empirically, because the nature of the alternate or true "state of affairs" is unknown, but generally the lower the level of β, the higher the level of α. To reduce the probability of making a Type II error, we need to increase the alpha level (from .01 to .05, for example).

The problem is one of balance. By lowering α, we increase β. In preliminary or pilot research, the rule of thumb is to use a higher level of significance like $\alpha = .05$. In a well developed field, with lots of experimental control, we tend to use stricter, lower, significance levels like $\alpha = .01$.

In summary:

As α decreases, p(Type I) decreases and p(Type II) increases.
As α increases, p(Type I) increases and p(Type II) decreases.

A Type II error is made when the null is false but we do not reject it. This is not a desirable state of affairs since the whole reason for testing hypotheses is to reject the null when it is false. Fortunately, there is another way to reduce β, the probability of making a Type II error.

Increasing the number of observations in the sample(s) reduces β. With larger samples, the standard error (the denominator of the Z formula) tends to be reduced. This is because there is less variability in larger samples. When the denominator is reduced, the Z value tends to increase and so it is more likely that the obtained Z value will be larger than the critical Z value. In other words, by increasing sample size, we increase the probability of rejecting a false null, which is what we are trying to do. When a test of significance has a high probability of rejecting false null hypotheses, it is said to be a **powerful** test.

Power

If β is the probability of *not rejecting a false null*, then clearly, the probability of *rejecting a false null* must be $1 - \beta$. The probability of rejecting a false null hypothesis is called **power**. Power, then, is the capability of our test to reject the null when it should be rejected. In practice, neither β nor power $(1 - \beta)$ are directly computed. However, increasing the α level (from .01 to .05, for example) or increasing the sample size, increases power.

The following figure illustrates what can happen when we test hypotheses about population parameters.

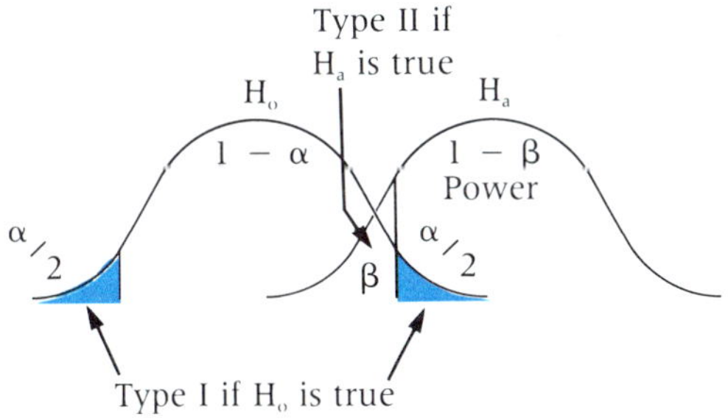

The curve on the left represents the random sampling distribution when the null hypothesis is true. The curve on the right represents one of many possible alternate hypotheses. You can see that if the null is true, occasionally we will obtain an outcome in the region of rejection that leads us to reject the null and make a Type I error. This will occur α of the time. Most of the time $(1 - \alpha)$, however, we will not reject a true null and so we make a correct decision. Look at the curve on the right. Here we see what could happen if the null is false and the alternate is true. If our sample outcome lies in the "region of acceptance," we will not reject the null even though it is false. This will occur β of the time and we will make a Type II error. However, $1 - \beta$ of the time, when the alternate is true, we will get a sample outcome in the critical region, reject the null, and make a correct decision. This is power.

We can summarize this as follows:

	True State of Affairs	
Decision	H_0 is True	H_0 is False
reject H_0	Type I $p = \alpha$	Correct Decision $p = 1 - \beta$ (Power)
do not reject H_0	Correct Decision $p = 1 - \alpha$	Type II $p = \beta$

Assumptions Underlying Inference with the Normal Curve

Inferential statistical procedures using the normal curve as a model assume certain conditions.

1. The sampling distribution of the statistic is normally distributed.
2. The sample observations have been randomly selected.
3. With two samples, observations have been independently assigned.

We know from the Central Limit Theorem that the sampling distribution will be normally distributed if the population is normal or if the sample(s) is large enough.

The assumption of random sampling and independent assignment is an important one. In research, we are careful not to violate this assumption. Random sampling and independent assignment of subjects to groups allows us to use very powerful statistical techniques. When these assumptions have been violated, other procedures are available but these are not, usually, as powerful as the techniques we have been studying.

SUMMARY OF FORMULAS

	Hypothesis Testing	Interval Estimation
The Mean	$Z = \dfrac{\bar{X} - \mu_{\bar{x}}}{\sigma_{\bar{x}}}$	$\mu = \bar{X} \pm Z\sigma_{\bar{x}}$
The Difference	$Z = \dfrac{\bar{X}_1 - \bar{X}_2 - (\mu_1 - \mu_2)}{\sigma_{\bar{x}_1 - \bar{x}_2}}$	$\mu_1 - \mu_2 = (\bar{X}_1 - \bar{X}_2) \pm Z\sigma_{\bar{x}_1 - \bar{x}_2}$
The Proportion	$Z = \dfrac{p - P}{\sigma_p}$	$P = p \pm Z\sigma_p$

Formulas for the Standard Error

The Difference	$\sigma_{\bar{x}_1 - \bar{x}_2} = \sqrt{\dfrac{\sigma_1^2}{n_1} + \dfrac{\sigma_2^2}{n_2}}$
The Mean	$\sigma_{\bar{x}} = \sigma/\sqrt{n}$
The Proportion	$\sigma_p = \sqrt{PQ/n}$

EXERCISES

1. The graduating averages of high school students in a large city are normally distributed with a mean of 79 and a standard deviation of 14. A random sample of 25 students from one of the schools has a mean of 85. Use a two-tailed test at $\alpha = .01$ to test the hypothesis that students from this school are at par with the city average.

2. Using the data from question 1, set up the 95% and 99% confidence intervals for the sample of 25 students.

3. Clare Voyent claims to have ESP. An investigator tests her with a deck of 52 cards, half red and half black. Clare guesses whether successive cards, turned over out of her sight, are red or black. She guesses correctly 29 times. Use a one-tailed test at $\alpha = .05$ to find out if Clare is doing significantly better than chance.

4. A seat belt manufacturer claims that his product has a mean breaking strength of 250 kg with a standard deviation of 3.5 kg. You select a random sample of 49 of his seat belts and compute the mean breaking strength of your sample to be 245 kg. Test the manufacturer's claim at $\alpha = .05$.

5. A realtor claims that more than 75% of retired couples prefer apartment living to single unit housing. A random sample of 100 couples shows that 81 prefer apartment living. Test the claim of the realtor at the 5% level of significance. (Hint: The realtor's claim that more than 75% prefer apartment living, corresponds to the alternate hypothesis. The null will state that 75% prefer apartment living.)

6. A normal population has a standard deviation of 20. A researcher collects a random sample of 100 scores and computes a mean of 48. Between what two values will the researcher be 99% confident that the population mean lies?

7. A toothpaste company claims that 70% of Canadian dentists use its toothpaste. A random sample of 50 dentists finds that 30 of these use this brand of toothpaste. Test the null hypothesis that the true proportion is 70% at $\alpha = .01$.

8. A total of 173 students were polled for an upcoming election and asked to state their preference for one of two candidates. Candidate A received 98 votes and B got 75. Test at $\alpha = .05$ whether Candidate A is ahead of Candidate B. (Hint: If A is not ahead of B, then what proportion of the total votes would we expect A to get?).

CHAPTER 10

Inference with the t-Distribution

To use the normal curve for estimating the value of population parameters and testing hypotheses about population parameters, a standard score called a Z-score must be determined. To test hypotheses about the value of μ, for example, we used the following to determine Z:

$$Z = \frac{\overline{X} - \mu}{\sigma_{\overline{X}}}$$

This formulation follows the general format for testing the null hypothesis:

$$\text{Standard Score} = \frac{\text{value of statistic} - \text{hypothesized value of the parameter}}{\text{standard deviation of the distribution of the statistic}}$$

To estimate the value of μ from sample data, we used

$$\mu = \overline{X} \pm Z\,\sigma_{\overline{X}}$$

The general formulation, then, for estimating the location of parameter values from samples is

Estimate of parameter = statistic ± standard score (standard deviation of the distribution of the statistic)

Let's look more closely at the two formulas. $\overline{X}$ is determined from the data that we collect. μ is the hypothesized value of the population mean or, in the case of confidence interval estimation, it is unknown. $\sigma_{\overline{X}}$ is the standard deviation of the sampling distribution of means calculated for all possible samples drawn at random from the population of raw scores. We called this the *standard error*. The standard error is determined by its relationship to σ: the standard deviation of the raw score population.

Here is the problem! *Since we don't know enough about the original population to know its mean, how likely are we to know its standard deviation?* Practically speaking, if we don't know μ, we probably don't know σ either!

To determine σ, we would have to look at all the raw scores in the population. If we could we would also know the mean of that population and wouldn't need to make inferences about what it might be. Instead, we would use descriptive statistical techniques, not inferential techniques.

When we do not know the value of σ, we cannot calculate the standard error or use Z to estimate where μ may lie or to test hypotheses about the value of μ.

In most research we do not know the value of σ, and the normal curve cannot be used as a model for making inferences about population parameters. And, as you have probably figured out already, we use a different curve or distribution as our model when we do not know the value of σ.

The t-Distribution and Unbiased Estimates

You will recall that the normal curve is the sampling distribution of the Z statistic. Similarly, the **t-distribution** is the sampling distribution of the t statistic. We use the t-distribution as a model when we do not know the value of the population standard deviation and must estimate its value from our sample data. We use lower case "**s**" to indicate that we are estimating the value of the population standard deviation. In summary:

Z-score: The deviation of a particular value from the mean of its distribution expressed in relationship to the standard deviation of that distribution.

t-score: The deviation of a particular value from the mean of its distribution expressed in relationship to an **estimate** of the standard deviation of that distribution.

The estimate of the population standard deviation used in the t-ratio is the square root of the **unbiased** estimate of the population variance.

An unbiased estimate of a parameter is one which, on the average, will equal exactly the parameter being estimated. To determine whether an estimate is unbiased:

Step 1. Draw all possible samples of some fixed size from the population.

Step 2. Calculate a statistic for each sample.

Step 3. Compute the mean of the statistics.

If this mean has the same value as the parameter to which it corresponds, then the estiminate is unbiased. Otherwise, it is a biased estimate.

Earlier, you learned that the mean of the sampling distribution of means ($\mu_{\bar{X}}$) is exactly equal to the mean of the population from which the samples were drawn (μ). In other words, the mean of all the sample means ($\bar{X}$'s) exactly equals the population mean. $\bar{X}$, then, is an *unbiased* estimate of μ.

Consider the statistic, **sample variance** (we'll call this S^2). The formula for calculating the variance of a sample is

$$S^2 = \frac{\Sigma(X - \bar{X})^2}{n} \qquad \text{Sample Variance}$$

If we calculated this S^2 statistic for all possible samples drawn from some population, and computed the mean of all the sample variances, we would find that this mean value is consistently smaller than σ^2. Although the mean of all the sample means will equal the mean of the population, the mean of all the sample variances *will not equal* the population variance. For this reason, we say that sample variance is a **biased estimate** of σ^2. It is biased because it always underestimates σ^2.

Fortunately, we have an arithmetic adjustment to correct this problem. Rather than calculating sample variance by the formula above, we can calculate s^2 as an unbiased estimate of σ^2.

$$s^2 = \frac{\Sigma(X - \overline{X})^2}{n - 1}$$ Unbiased Estimate of Population Variance

As you can see, s^2 uses a denominator of $n - 1$, instead of n. This results in a somewhat larger quotient. Now, if we were to draw all possible samples from a population, calculate s^2 for each sample, and compute the mean of all the s^2's, we would have a value which exactly equals σ^2. This formula, then, produces an unbiased estimate of σ^2 because its expected value is σ^2.

Recall that the t-ratio is the deviation of a value, from the mean of its distribution, expressed in relationship to an estimate of the standard deviation of that distribution. This estimate is always based on the unbiased estimate of the population variance. Some estimates used in inference with the t-distribution are

Estimate	Parameter
$\overline{X}$	μ
s	σ
s^2	σ^2
$s_{\overline{X}}$	$\sigma_{\overline{X}}$
$s_{\overline{X}_1 - \overline{X}_2}$	$\sigma_{\overline{X}_1 - \overline{X}_2}$

Relationship between the Normal and the t-Distribution

Both the normal and the t-distribution:

1. are symmetrical
2. are unimodal
3. have means of zero

Recall that the Z equivalent for a sample mean is

$$Z = \frac{\overline{X} - \mu_{\overline{X}}}{\sigma_{\overline{X}}}$$

Similarly, the t equivalent for a sample mean is

$$t = \frac{\overline{X} - \mu_{\overline{X}}}{s_{\overline{X}}}$$

where $s_{\overline{X}}$ is the estimate of the standard error.

If you drew all possible samples of some fixed size from a population, calculated Z-scores and t-scores for each sample mean, placed the Z-scores and t-scores in separate frequency distributions, the Z's would follow a normal distribution and the t's would follow a t-distribution.

The difference between the normal and the t-distribution is mainly in the tails. The t-distribution tends to have more area in the tails than does the normal distribution. The Z- distribution is normal in shape because μ and σ are fixed and the sample means are normally distributed or close to it (remember the Central Limit Theorem). Since μ and σ do not change, the distribution of Z values follows the shape of the distribution of the sample means: they are the only values that vary. We know that the distribution of sample means is close to normal, regardless of the shape of the original population when sample size is reasonably large.

The t-distribution, on the other hand, departs from normality in a regular fashion even though the sample means are normally distributed. μ is fixed but the estimate of $\sigma_{\bar{x}}$ (i.e., $s_{\bar{x}}$) is determined from the sample data and therefore, is a variable differing from sample to sample. The variability of $s_{\bar{x}}$ is directly affected by sample size; smaller samples produce greater variability.

The exact shape of the t-distribution depends on **degrees of freedom (df).** Recall that the normal curve is a family of curves that has been standardized to the Standard Normal with Z-scores on the abscissa and relative frequency on the ordinate. The t-distribution is also a family of curves the shape of each depending on degrees of freedom. Unless degrees of freedom are infinitely large, the shape of the t-distribution will not be normal. Now, with numerous degrees of freedom, the t-distribution does look close to normal, but with very few degrees of freedom the t-distribution is leptokurtic with respect to the normal curve. Does anyone remember what that means? You can see from the examples below that the t-distribution has more area in the tails than the normal distribution.

Examples of the t-distribution

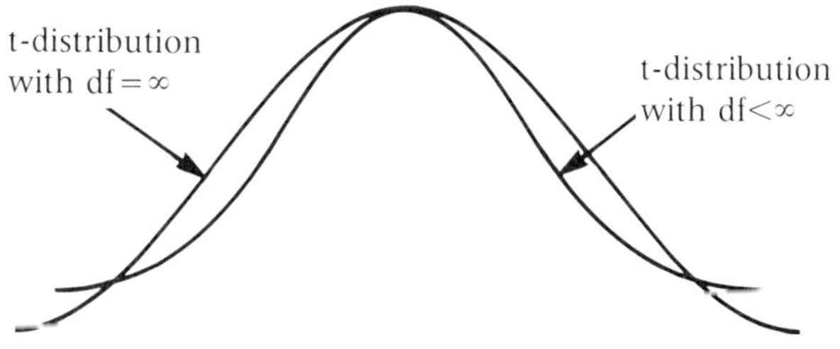

Degrees of freedom are related to, but not identical with, sample size. When degrees of freedom are infinitely large, the shape of the t-distribution is normal and the value of the t-ratio equals exactly the value of the Z-ratio.

In such cases, there is no *mathematical* distinction between calculating Z and calculating t.

When degrees of freedom are less than infinity, the t-distribution has more area in the tails than does the normal distribution, and the value of the t-ratio will not equal the value of the Z-ratio. When testing the null hypothesis, somewhat larger t values are required to reject the null hypothesis whenever degrees of freedom are less than infinity. The critical values of t are larger than the critical values of Z and depend on degrees of freedom.

Degrees of Freedom when Estimating Parameters

Suppose you overheard a conversation, about your performance on a test, between your statistics instructor and her friend Clive. It included the following information: 1. You scored 150 points on your statistics exam. 2. You missed 50 points. 3. A perfect score on the test was 200. You have three pieces of information: your score, the points you missed, and the total number of points possible. Actually, you could have deduced the third piece of information from the first two. Knowing that you scored 150 and missed 50 points is all you need to deduce that the total number of points on the test was 200. In fact, any two of the three values is all you need to obtain the last value. For example, if you heard that you lost 50 points on the test and there were 200 points available you could compute your score by subtraction.

In situations like this, statisticians say that two pieces of information fixed the third. They also say that two of the values are independent and one is dependent. Statisticians tend to state the same thing in as many ways as possible; many statisticians will also say that two values are free to vary and one is fixed.

Degrees of freedom, then, in any particular situation, are determined by the number of values free to vary. This concept is important when we estimate parameter values from sample data. When we estimate σ with our s formula, which we must do to determine the t-ratio for a single mean, we use the sample mean ($\overline{X}$) as an estimate of μ, and by doing so we use up one degree of freedom. In other words, calculating $\overline{X}$ and using it in our computation of s, fixes one value in that it is no longer free to vary.

This point is clear, if we consider the following distribution of scores. The first column provides the original raw scores, and the second column shows the deviation of each score from the mean of the distribution.

X	$(X - \bar{X})$
15	2
14	1
13	0
12	−1
11	−2
$\bar{X} = 13$	$\Sigma(X - \bar{X}) = 0$

You will recall that the sum of the positive deviations must exactly equal the sum of the negative deviations for deviations taken around the mean of any distribution. The mean of this X distribution is 13 and the sum of the negative deviations equals the sum of the positive deviations around 13. How many score values could we change without changing the value of the mean?

Let's change the first value from 15 to 20, the second from 14 to 17, the third from 13 to 10, and the fourth from 12 to 8. Before we go further let's determine the deviation scores. Remember, we cannot change the value of the mean.

X	$(X - \bar{X})$
20	7
17	4
10	−3
8	−5
?	___
$\bar{X} = 13$	$\Sigma(X - \bar{X}) = 0$

Keeping the mean at 13 then, we have a sum of 11 positive deviations [i.e., $(20 - 13) + (17 - 13) = 11$] and a sum of 8 negative deviations [i.e., $(10 - 13) + (8 - 13) = 8$]. Because the sum of the negative deviations must equal the sum of the positive deviations around any mean, our fifth score value must be 3 deviations below the mean. Our fifth value, then, can only be a 10. No other value will do! We have only 4 degrees of freedom. Once four values are changed the fifth value is fixed.

When we estimate the population standard deviation using the sample mean, we constrain the data and use up a degree of freedom. Degrees of freedom are determined by the number of constraints placed on our data by our calculations. When computing a t-score for the mean of a single sample, we estimate the population standard deviation from our sample data; in doing so we use up one degree of freedom. Degrees of freedom, then, are equal to $n - 1$, the number of values free to vary.

When computing a t-score for the difference between means, we estimate population standard deviation using the means of both samples; in doing so, we use up a degree of freedom for each mean calculated. Degrees of freedom, then, are equal to $(n_1 - 1) + (n_2 - 1)$.

Using the t-Distribution

Just as the normal curve may be used as a model when making inferences or testing hypotheses about the value of population parameters under certain conditions, so can the t-distribution be used as a model for such procedures under other conditions. The choice depends on whether or not the population standard deviation is known. If it is, the normal curve is the appropriate model: if it must be estimated, than t may be the appropriate distribution.

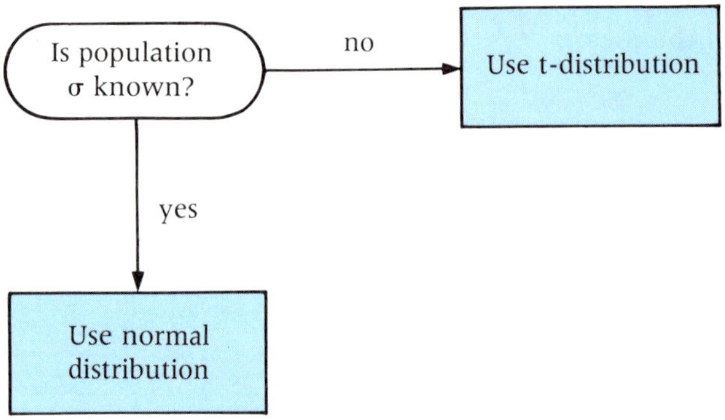

Confidence Intervals using the t-Distribution

The t-distribution is the sampling distribution of the t-statistic. Like Z, t is calculated for all possible samples of some fixed size drawn from a population. Setting up confidence intervals with the t-distribution is essentially the same as it was using the normal curve and Z.

Setting up the Confidence Interval for a Population Mean

You will recall that the normal curve formula for setting up a confidence interval to locate the population mean is

$$\mu = \overline{X} \pm Z_{crit}(\sigma_{\overline{X}}) \qquad \text{CI Normal Curve}$$

The formula for setting up a confidence interval to locate the population mean using the t-distribution is

$$\mu = \overline{X} \pm t_{crit}(s_{\overline{X}}) \qquad \text{CI t-Distribution}$$

Rather than using the standard error, we must use an estimate of the standard error. The critical value of t changes with the significance level chosen, as does the critical value of Z, and with the degrees of freedom, unlike Z. So, unless you are going to use the same sample size whenever you do research, I am afraid you are going to have to look up the critical value of t for each confidence interval you set up! Table 2, at the back of the book, lists critical values of t depending on degrees of freedom and alpha level.

Ready for an example? The chairperson of the Ski Jump Preparation Committee is in the process of making decisions about the facilities for the next Olympics. She is not sure how long the landing platform should be. She decides to set up the 95% Confidence Interval for the mean length required for landing by world class ski jumpers. (Don't ask me why she didn't confer with the chairperson of the previous Ski Jump Preparation Committee!). She selects a random sample of 25 world class ski jumpers and measures the distance required by each for landing. Here are her data:

Mean Distance: $\overline{X} = 125$	$n = 25$	$df = n - 1 = 24$
$s = 15$	$t_{.05} = 2.064$	

The appropriate formula to use is

$$\mu = \overline{X} \pm t_{.05}(s_{\overline{X}})$$

Notice that the estimate of the population standard deviation (i.e., "s") is given. You may recall that in Chapter 9 the standard error was computed by dividing the population standard deviation by the square root of the sample size. We do the same here to obtain the estimate of the standard error. We compute $s_{\overline{X}}$ by dividing s by $\sqrt{n}$.

$$s_{\overline{X}} = s/\sqrt{n}$$
$$s_{\overline{X}} = 15/\sqrt{25}$$
$$= 3.0$$

Now we can proceed to set up the 95% Confidence Interval for μ:

$$\mu = 125 \pm (2.064)(3.0)$$

Limits are: 118.81 to 131.19

Using the "official" format for presenting confidence interval limits:

$$C(118.81 \leq \mu \leq 131.19) = .95$$

The Chairperson of the Ski Jump Preparation Committee can claim, in the next meeting, that her confidence that the average world class ski jumper jumps distances from 118.81 to 131.19 meters is 95%. This may strike you as a ludicrous example, but I hope you get the idea!

Setting up the Confidence Interval for the Difference between Independent Population Means

Setting up confidence interval for the difference between population means follows the same general pattern. The appropriate formula is

$$\mu_1 - \mu_2 = (\overline{X}_1 - \overline{X}_2) \pm t_{crit}(s_{\overline{X}_1 - \overline{X}_2})$$

CI for the difference between Independent Means

The trickiest part of this procedure is determining the estimate of the standard error of the difference. One formula is

$$s_{\overline{X}_1 - \overline{X}_2} = \sqrt{s^2_{\overline{X}_1} + s^2_{\overline{X}_2}}$$

This formula is fine if we have already calculated the unbiased estimates of the variance for each sample. If we have not, however, the following computational formula is appropriate:

$$s_{\overline{X}_1 - \overline{X}_2} = \sqrt{\frac{\Sigma x_1^2 + \Sigma x_2^2}{n_1 + n_2 - 2}\left(\frac{1}{n_1} + \frac{1}{n_2}\right)}$$

Estimate of the Standard Error of the Difference when $n_1 \neq n_2$

This formula is appropriate when the two samples are not the same size. When $n_1 = n_2$, the formula simplifies somewhat to:

$$s_{\overline{X}_1 - \overline{X}_2} = \sqrt{\frac{\Sigma x_1^2 + \Sigma x_2^2}{n(n - 1)}}$$

Estimate of the Standard Error of the Difference when $n_1 = n_2$

Chas Chauvinist wonders what the difference might be between men and women on a test of creativity he has creatively created. He decided to use a random sample of 15 men and another of 15 women to set up the 99% CI for the difference. Unfortunately, the data from two of the female subjects were lost. Chas decides to continue with the data he has. These are the values he obtained.

Creativity Test Data

For the women: $\overline{X}_1 = 98$, $n_1 = 13$, $\Sigma x_1^2 = 30$

For the men: $\overline{X}_2 = 100$, $n_2 = 15$, $\Sigma x_2^2 = 22$

$df = n_1 + n_2 - 2 = 26$

$t_{.01} = \pm 2.779$

The first thing Chas has to do is to compute the estimate of the standard error of the difference. Since the sample sizes were not the same, he used

$$s_{\overline{X}_1 - \overline{X}_2} = \sqrt{\frac{\Sigma x_1^2 + \Sigma x_2^2}{n_1 + n_2 - 2}\left(\frac{1}{n_1} + \frac{1}{n_2}\right)}$$

$$= \sqrt{\frac{30 + 22}{13 + 15 - 2}\left(\frac{1}{13} + \frac{1}{15}\right)}$$

$$= \sqrt{\frac{52}{26}(0.15)} = 0.55$$

Now the 99% CI can be determined.

$$\mu_1 - \mu_2 = (\overline{X}_1 - \overline{X}_2) \pm t_{crit}(s_{\overline{X}_1 - \overline{X}_2})$$

$$= (98 - 100) \pm 2.779(0.55)$$

The limits are 0.47 to 3.53

Chas may say, with 99% confidence, that the mean difference between men and women on his test of creativity lies between 0.47 and 3.53.

To present these data formally:

$$C(0.47 \leq (\mu_1 - \mu_2) \leq 3.53) = .99$$

Hypothesis Testing with the t-Distribution

Testing hypotheses about the value of a population parameter(s), using the t-distribution, follows the same logic and procedure as it does with the normal curve. The general form of the t-ratio is similar to the general form of the Z-ratio.

$$t = \frac{\text{value of statistic} - \text{hypothesized value of parameter}}{\text{estimate of standard deviation of distribution of the statistic}}$$

If $\overline{X}$ is the statistic in question, then

$$t = \frac{\overline{X} - \mu_{\overline{X}}}{s_{\overline{X}}} \quad \text{t for a single mean}$$

If $\overline{X}_1 - \overline{X}_2$ is the statistic in question, then

$$t = \frac{\overline{X}_1 - \overline{X}_2}{s_{\overline{X}_1 - \overline{X}_2}} \quad \text{t for the difference between means}$$

Let's go ahead and test some hypotheses using the t-distribution.

Testing an Hypothesis about a Population Mean

A nation wide survey conducted in the 60's reported that young people listened to rock music an average of 3.5 hours per day. Let's hypothesize that the youth of today are not the "rockers of yesterday." We will randomly select a sample of twenty-five 16 year-olds and record the numbers of hours per day each spends listening to rock music over a period of 6 months. Let's use a one-tailed test to test our hypothesis at $\alpha = .05$.

H_0: $\mu = 3.5$ $\qquad$ $\overline{X} = 1.89$

H_a: $\mu < 3.5$ $\qquad$ $s = 1.72$

$df = 25 - 1 = 24$ $\qquad$ $t_{.05} = -1.71$

To use our t formula, we first need to determine the estimate of the standard error.

$$s_{\overline{X}} = s/\sqrt{n} = 1.72/\sqrt{25} = 0.34$$

$$t = \frac{1.89 - 3.5}{0.34} = -4.68$$

Since our obtained t value is larger numerically than the critical value for a one-tailed test with 24 df, we reject the null hypothesis. We conclude that "the youth today don't rock like they used to."

Summary of t-Test for a Single Mean

Hypotheses

H_o: μ = specified value

H_a: $\mu \neq, >, <$ specified value

Assumptions

1. Subjects are randomly selected
2. Population distribution is normal

Decision Rules

df = n − 1

If $t_{obt} \geq t_{crit}$, reject the H_o

If $t_{obt} < t_{crit}$, do not reject H_o

Formula

$$t = \frac{\overline{X} - \mu}{\sqrt{\dfrac{\Sigma X^2 - (\Sigma X)^2/n}{n(n-1)}}}$$

Testing an Hypothesis about the Difference between Independent Population Means

A t-test for the difference between two independent means follows the same general logic and procedure as with the normal curve test. Let's consider the following example.

A psychology professor is interested in knowing whether there is a difference in final exam performance between students who take their introductory psychology course under a personalized system of instruction (PSI), and those who study under more conventional, lecture-discussion, format (L). She selects a random sample of 30 students and randomly assigns 15 of them to each of two groups. One group studies under PSI conditions and the other under L conditions. Unfortunately, two of the PSI students dropped out of school before they wrote the final exam. Nevertheless, she goes ahead and tests her hypothesis.

$$H_o: \mu_1 - \mu_2 = 0$$
$$H_a: \mu_1 - \mu_2 \neq 0$$

L condition	$\bar{X}_1 = 75.0$	$\Sigma x_1^2 = 2016$	$n_1 = 15$
PSI condition	$\bar{X}_2 = 86.0$	$\Sigma x_2^2 = 1200$	$n_2 = 13$

$$df = n_1 + n_2 - 2 = 26$$
$$t_{.05} = \pm 2.056$$

$$t = \frac{\bar{X}_1 - \bar{X}_2}{\sqrt{\left(\frac{\Sigma x_1^2 + \Sigma x_2^2}{n_1 + n_2 - 2}\right)\left(\frac{1}{n_1} + \frac{1}{n_2}\right)}}$$

$$= \frac{75 - 86}{\sqrt{\left(\frac{2016 + 1200}{15 + 13 - 2}\right)\left(\frac{1}{15} + \frac{1}{13}\right)}} = -2.61$$

Since the obtained value of t is numerically larger than the critical value, she may reject the null hypothesis. She has evidence that there is a significant difference in performance under PSI versus L conditions.

Summary of t-Test for Independent Means

Hypotheses

H_o: $= \mu_1 = \mu_2$
H_a: $\mu_1 \neq \mu_2$ or $\mu_1 < \mu_2$ or $\mu_1 > \mu_2$

Assumptions

1. Subjects are randomly selected and independently assigned to groups.
2. Population variances are homogenous.
3. Population distributions are normal.

Decision Rules

$df = n_1 + n_2 - 2$
If $t_{obt} \geqslant t_{crit}$, reject H_o
If $t_{obt} < t_{crit}$, do not reject H_o

Formula

$$t = \frac{\bar{X}_1 - \bar{X}_2}{\sqrt{\left(\frac{\Sigma x_1^2 + \Sigma x_2^2}{n_1 + n_2 - 2}\right)\left(\frac{1}{n_1} + \frac{1}{n_2}\right)}}$$

Assumptions Underlying Inference with the t-Distribution

Homogeneity of variance: Equal variances in populations.

The use of the t-distribution in making inferences about population parameters assumes certain conditions. One very important assumption is that the observations were randomly selected from the population(s) of interest. Second, the estimate of the standard error is based on an unbiased estimate of the population variance. Third, the sampling distribution of the statistic (i.e., the sample mean or difference between means) must be normally distributed. You will recall that this will be true (according to the Central Limit Theorem) as long as either the population is normal or the sample size(s) is large enough. A further assumption with t-tests for the difference between means, is that the populations from which the two samples were collected have equal variability. This is called the assumption of **homogeneity of variance**. We want to know if we are sampling from distributions with different means, not different variances. It will probably occur to you that, in actual research, it is unlikely we will know whether the populations are equally variable or not. Not to worry! The t-test is quite insensitive to violations of this assumption as long as our samples sizes are large enough and *approximately equal.*

For the t-test for the difference between two independent means we have another very important assumption. It is assumed that the observations have not only been randomly selected but that they have been *assigned randomly* to the two samples. In other words, the samples are independently formed. The observations in the first sample must not be correlated or dependent on observations in the second sample.

There are many occasions during research where we do not have independent samples; for these situations we have another procedure for making inferences with the t-distribution. When samples are dependent or correlated we will use the t-test for Dependent Samples.

Dependent or Correlated Samples

Probably the major grievance of researchers in the "softer" sciences is the extreme variability of the data. People, for example, do not behave like molecules. Chemists are fortunate. For the most part, when they repeat experiments, they get identical results every time. Research in psychology and related disciplines is necessarily much less precise.

All statistical inference is really asking the same question. Is our obtained outcome likely to be due to chance? When we obtain a sample outcome that seems highly unlikely to have occurred if the assumptions of the null are true, we say we have a **significant** result. It seems more reasonable to conclude that the null is not true than to conclude that we got such an unlikely result by chance alone. In other words, the subjects in our experiment did not behave the way we would expect them to behave if the assumptions of the null were correct. Let's look more closely at the behavior of a subject. Whatever we are dealing with, an animal, an insect or a person, a score for any single subject in a research study reflects at least three factors:

1. Random error.
2. The effects of the experimental variable.
3. The ability of the subject to do the task.

Random error is one of the factors that makes behavioral research so imprecise (I call this the slop factor). We do our best to control this factor by using good experimental techniques.

The effects of the experimental variable on our subjects' behavior is the factor of interest in any experiment. We would like our experimental variable to have a big effect on the subjects' performance.

The third factor, ability, is often the most significant factor contributing to the scores and their variability in research in psychology and related disciplines. Many researchers try to reduce the effects of this factor by using subjects who do not differ much in ability. Many animals, for example, are very close in ability and consequently reveal less variable behavior than do people. Research with people, however, must deal with the ability factor constantly.

Random assignment of subjects to groups is a powerful technique which allows us to assume *initial* "between-group" equality. In other words, if we have randomly assigned subjects to each group in our experiment, we may assume that the groups do not differ in average ability. If, after the introduction of some experimental variable, the groups perform differently, then we may conclude that this difference is due to the variable, not to differences in initial ability. Random assignment of subjects to groups is not the only way to control between group differences in initial ability. The use of **correlated** or **dependent samples** is another way of reducing the effect of subject differences on the scores.

Within-Subjects Designs

Many studies of the effects of an experimental variable use the same subjects in both the **treatment** (experimental) group and in the **control** group. Measuring performance of subjects before and after they receive some in-

Treatment group: The group in an experiment that is exposed to the independent variable.

Control group: The group that is not exposed to the independent variable.

structional training is an example of a **within-subjects design**. Rather than measuring performance between two groups, one of which receives the training and the other of which does not (a **between-subjects design**), we measure performance of the same subjects under no training and training conditions. This reduces the between-subjects variability because a subject will behave more like himself than anyone else. The effect of this, then, is that the two groups do not differ in initial ability.

Matched-Groups Designs

Another way of reducing between-subject variability is to **match** subjects on some variable known to be related to the performance measure (usually the dependent variable). Giving subjects a practice-task similar to what they will be doing in the experiment, for example, would allow us to match subjects on initial ability and assign one subject to our control group and another of equal ability to our experimental group. In this way, we would end up with groups of equal initial ability.

When to Use Dependent-Groups Designs

Error term: The denominator of the significance-test ratio.

Both the within-subject and the matched-groups techniques serve to reduce the differences in initial ability between groups and, therefore, let us better estimate the effects of the experimental variable on performance. Arithmetically speaking, these procedures reduce the size of the denominator (**error term**) in the t-test. The standard error will be smaller because there is less variability between subjects. With a smaller denominator, the t-ratio will tend to be larger. The dependent samples approach, therefore, may be a more *powerful* test of significance of an experimental variable. You will recall that power is the term we use to describe the ability of the test to reject a false null. You might now ask, "why don't we always do dependent sample research?" Clearly if the t-ratio tends to be larger with a dependent sample approach, then we are more likely to reject the null... right?... well, not necessarily!

Nothing is so simple in statistics. It is true that the t-ratio for dependent samples tends to be larger than for independent samples. However, you will recall that the critical value of t, to which we compare our obtained value, depends on degrees of freedom. As degrees of freedom get larger our critical value gets smaller. Remember that for a t-test for two independent samples, the degrees of freedom are $n_1 + n_2 - 2$. With dependent samples degrees of freedom are $n_p - 1$ (number of *pairs* of scores $- 1$).

For example, if we did a study with two groups, ten observations in each group, we would have 18 df for an independent t-test but *only 9*df for a dependent t-test. Why this is so will become clear to you shortly. But for now realize that, while use of a dependent sample approach may increase the value of the obtained t-score, the critical value of t needed for rejection of the null is also increased. On the one hand, we increase the power of our test by reducing the error term but, on the other hand, we decrease power by losing degrees of freedom. This is why we do not always use a dependent sample approach.

Dependent-groups designs are very popular in psychology. In most cases, the effect of between-subject differences is the most significant source of error, and the dependent-groups approach is, overall, a more sensitive test of the null in spite of the loss of degrees of freedom. The reduction of the error term, the denominator of the ratio, usually has a greater impact on the outcome of the statistical procedure than does the loss of degrees of freedom.

There are situations, however, where within-subject designs should not be used. Sometimes the performance of the control task carries over to the experimental task. Measuring performance on a problem-solving task before and after instruction on some problem-solving strategy would be an example of such a case. The practice effects of working on the problems first under control conditions may carry over to the experimental condition and thus, mask the effects of the independent variable (i.e., the training). There are other conditions where a within-subjects design may not be desirable, but in many cases this approach is a useful and powerful one.

Matched-groups designs avoid the problem of carry over effects because they use different subjects in each group, matched on some initial measure. It is critical, with this approach, that the matching variable be an important contributing factor to task performance. If it is not, then, the t-test will be less powerful than an independent t-test because of the loss of degrees of freedom.

Generally speaking, a matched-groups design will be more powerful than an independent-groups design when the variable used to equate subjects is strongly related to performance on the task of interest. If we match subjects on some variable unrelated to performance on the task we are interested in, then we will have a less powerful test of the null hypothesis. For example, say we are interested in the effects of reward on the arithmetic performance of children. We decide to use two groups of children and have them do a whole series of arithmetic problems. One group receives a treat for each correct answer; the other group receives nothing. We count the number of problems solved, by each child, at the end of the session. Suppose that, for some obscure reason, we believe height may be an important contributing factor to task performance. So, we measure the heights of a group of children and find 20 pairs of children of the same height. We assign one child of each matched pair to the control (no reward) condition and the other child of the same height to the experimental condition (reward). This, somewhat silly, example illustrates a *weak test* of the null hypothesis. Our matching variable is probably unrelated to task performance, so we have not reduced the size of the error term in the t-ratio. Our obtained t-value, then, will be smaller and we have many fewer degrees of freedom (df = 19) than if we had not matched subjects (df = 38). Matched-groups designs can be powerful but we must be careful in selecting our matching variable.

If we have matched subjects or used the same subjects under both control and experimental conditions, we will run a dependent or correlated t-test on our data.

t-Test for the Difference between two Dependent Samples

One of the formulas you were given for the standard error used in the denominator of the t-ratio for the difference between means was:

$$s_{\bar{x}_1 - \bar{x}_2} = \sqrt{s^2_{\bar{x}_1} + s^2_{\bar{x}_2}}$$

Actually, this is not the case!

The real formula for any t-test of the difference whether it be a dependent-groups or an independent-groups test is:

$$s_{\bar{x}_1 - \bar{x}} = \sqrt{s^2_{\bar{x}_1} + s^2_{\bar{x}_2} - 2\rho s_{\bar{x}_1} s_{\bar{x}_2}}$$

You will appreciate why I didn't show you this earlier! Consider an independent t-test for the difference. This test requires that subjects be independently assigned to groups. If this is done, the performance of the two groups would not be expected to be correlated in any way. A measure of correlation which we will discuss later is symbolized by ρ. With independent samples we expect no correlation between the populations from which the groups were selected ($\rho = 0$). For this reason, we do not include the third term under the square root sign because it is assumed to be zero when samples are independent.

When sample are dependent, however, the correlation will not be zero and the full form of the standard error formula must be used. This formula requires that we determine (actually we estimate) the correlation before we can calculate the standard error. If you do not have a computer handy which will do this for you in no time at all, there is a convenient technique for computing the t-ratio for dependent samples. This technique is called the **direct difference method**. The t-ratio follows the same general format we have been using:

$$t = \frac{\overline{D} - \mu_{\overline{D}}}{s_{\overline{D}}}$$ t-ratio for Dependent Means: Direct Difference Method

The differences between the pair of measures for each subject are the data. We take the mean difference between our two groups (i.e., $\overline{D}$) and subtract from that the hypothesized mean difference (i.e., $\mu_{\overline{D}}$) and divide this by the standard error of the difference (i.e., $s_{\overline{D}}$). As with any two sample t-test, the hypothesized difference stated by the null is zero and so we often leave out the last term.

Ready for an example? An administrator at the University of Toronto is interested in the effects of a promotional film on student attitude toward large classes. He assumes that the film will produce more positive attitudes. He decides to measure the attitudes of a random sample of students before and after they view the film. Since the same subjects are tested before and after, he will analyze his data with a dependent t-test. Here are his data:

H_o: $\mu_1 = \mu_2$ $\alpha = .01$ $df = 9$

H_a: $\mu_1 < \mu_2$ $t_{.01} = -2.82$

Attitude Before Film	Attitude After Film	D	D^2
25	28	−3	9
23	19	4	16
30	34	−4	16
7	10	−3	9
3	6	−3	9
22	26	−4	16
12	13	−1	1
30	47	−17	289
5	16	−11	121
14	9	5	25
		$\Sigma D = -37$	$\Sigma D^2 = 511$

$$s_{\overline{D}} = \sqrt{\frac{\Sigma D^2 - (\Sigma D)^2/n}{n(n-1)}} = \sqrt{\frac{511 - (-37)^2/10}{10(9)}} = 2.04$$

$$t = \frac{\overline{D}}{s_{\overline{D}}} = \frac{(-37)/10}{2.04} = \frac{-3.7}{2.04} = -1.81$$

Since the obtained t value does not lie in the critical region, the null is not rejected. The administrator must conclude that he has no evidence that the students attitudes toward large classes were more positive after viewing the film.

The direct difference method uses difference scores to compute the standard error (denominator) and the mean difference between the groups to compute the numerator of the t-ratio. This technique takes into account the expected correlation.

The example demonstrates why degrees of freedom for a dependent t-test is "number of pairs of scores minus one." We used the ten differences as our data for calculating the t-ratio. Recall that any standard error is the square root of the quotient obtained by dividing the sum of squares by the appropriate degrees of freedom. The sum of squares is the sum of the squared deviations of the scores from the mean of the distribution. The standard error for the dependent samples t-test uses the *mean difference* in the computation of the sum of squares. In our example we had ten differences and we used the mean difference in calculating the sum of squares. Only nine of these differences are free to vary since the mean difference is fixed. Statistically we are operating as if we had only one group of scores: the difference scores. Therefore, the degrees of freedom are equal to the number of difference scores minus one.

Summary of t-Test for Dependent Means

Hypotheses

H_o: $= \mu_1 = \mu_2$

H_a: $\mu_1 \neq \mu_2$ or $\mu_1 < \mu_2$ or $\mu_1 > \mu_2$

Assumptions

1. Subjects are randomly selected.
2. Population distributions are normal.
3. Population variances are homogeneous.
4. Repeated measures or matched subjects are used.

Decision Rules

$df = n_{pairs} - 1$

If $t_{obt} \geq t_{crit}$, reject H_o. t_{obt} is greater than or = to t_{crit}

If $t_{obt} < t_{crit}$, do not reject H_o. t_{obt} is less than t_{crit}

Formula

$$t = \frac{\overline{D}}{\sqrt{\dfrac{\Sigma D^2 - (\Sigma D)^2/n}{n(n-1)}}}$$

SUMMARY OF FORMULAS

	Hypothesis Testing	Interval Estimation
The Mean	$t = \frac{\bar{X} - \mu_{\bar{X}}}{s_{\bar{X}}}$	$\mu = \bar{X} \pm t_{crit}(s_{\bar{X}})$
The Difference Between Independent Means	$t = \frac{\bar{X}_1 - \bar{X}_2 - (\mu_1 - \mu_2)}{s_{\bar{X}_1 - \bar{X}_2}}$	$\mu_1 - \mu_2 = (\bar{X}_1 - \bar{X}_2) \pm t_{crit}(s_{\bar{X}_1 - \bar{X}_2})$
The Difference Between Dependent Means	$t = \frac{\bar{D} - \mu_{\bar{D}}}{s_{\bar{D}}}$	$\mu_{\bar{D}} = \bar{D} \pm t_{crit}(s_{\bar{D}})$

Computational Formulas for the Estimates of:

Population Standard Deviation

$$s = \sqrt{\frac{\Sigma X^2 - (\Sigma X)^2/n}{n - 1}}$$

Standard Error of the Mean

$$s_{\bar{X}} = s/\sqrt{n}$$

$$= \sqrt{\frac{\Sigma X^2 - (\Sigma X)^2/n}{n(n - 1)}}$$

Standard Error of the Difference

$$s_{\bar{X}_1 - \bar{X}_2} = \sqrt{s^2_{\bar{X}_1} + s^2_{\bar{X}_2}}$$

Independent Samples

$$= \sqrt{\left(\frac{\Sigma x_1^2 + \Sigma x_2^2}{n_1 + n_2 - 2}\right)\left(\frac{1}{n_1} + \frac{1}{n^2}\right)} \qquad n_1 \neq n_2$$

$$= \sqrt{\frac{\Sigma x_1^2 + \Sigma x_2^2}{n(n - 1)}} \qquad n_1 = n_2$$

Dependent Samples

$$s_{\bar{D}} = \sqrt{\frac{\Sigma D^2 - (\Sigma D)^2/n}{n(n - 1)}}$$

EXERCISES

1. A group of 23 subjects was given a list of 30 one-word anagrams to solve. Half the solution words were emotionally neutral in meaning and half were unpleasant. Which t-test should you run to determine if affective or emotional meaning of the solution word has any effect on the time to solve the anagrams?

2. A psychologist wanted to find out whether the development of the ability to formulate abstract concepts could be hastened by special training. She selected 10 sets of 5-year old identical twins to be her subjects. The twenty children were given 25 conceptual problems to solve, and it was found that none of them could offer any correct solutions. Following this initial test, one twin in each set was given 30 minute training sessions every day for three weeks. Six months after the training was complete, all 20 children were again given the same set of 25 conceptual problems they had been given earlier. The measure was the number of correct solutions on the second test. Assume that all subjects improved on the second testing. Which t-test should you use to determine if the special training had any additional effect?

3. A child psychologist is interested in the effects of 2 teaching methods in a learning task. She believes, however, that intelligence and gender may be involved. She randomly selects 20 male children all with IQs over 120. She randomly divides them into two groups, one of which is taught by one method and the other by the second method. She measures amount of time to learn required by each child. Which t-test should be used to determine if teaching method makes a difference in performance?

4. Indicate the degrees of freedom and the critical value of t at $\alpha = .05$ for the following:

   a) two-tailed t-test for Independent means
   $n_1 = 18 \quad n_2 = 12$

   b) two-tailed t-test for Dependent means
   $n_1 = 16 \quad n_2 = 16$

5. Sixteen students randomly selected from an Introductory Psychology class are assigned randomly to two experimental conditions. Both groups are given the same 25-item multiple choice test on class material. One group (Condition A) is given hints on taking multiple choice tests; the other group (Condition B) is not. At $\alpha = .01$, run a t-test on the data below. Indicate your null and alternate hypothesis and the critical t value. What will be your statistical decision?

Condition A	Condition B
12	16
17	14
15	18
13	19
11	17
10	13
16	11
12	10

6. Fifteen randomly selected students from the University of Toronto are given two forms of an attitude questionnaire one before and the other after listening to a speaker discuss the separation of Quebec from the rest of Canada. Run a

t-test on the data below to determine if the speaker had an influence on student attitude regarding separation. Indicate your null and alternate hypothesis and the critical value of t at $\alpha = .01$. What is your statistical decision?

Attitude Score Before Speaker	Attitude Score After Speaker
13	22
17	14
14	23
17	21
23	20
20	26
13	14
25	27
24	20
18	21
17	15
15	29
21	30
19	22
28	28

7. An experimenter wants to know if reward has any effect on the learning rate of gerbils. He collected data in the form of number of trials to perfect performance for two randomly selected groups of 10 gerbils each. The reward group received a "gerbil treat" at the end of each trial but the control group received nothing. Unfortunately, four of the gerbils in the control group contracted "gerbil fever" in their home cages and were unable to finish the experiment. Nevertheless, he did a t-test on the remaining data. You do one too on the data below at $\alpha = .05$. Indicate your null, alternate and critical value of t. What is your statistical decision?

Mean Performance	
Control	Reward
12	6
10	5
12	6
20	4
10	3
8	8
	10
	6
	4
	8

8. Set up the 95% CI for the difference between means for the following data:

$\bar{X}_1 = 25$	$\bar{X}_2 = 20$
$n_1 = 15$	$n_2 = 10$
$\Sigma x_1^2 = 225$	$\Sigma x_2^2 = 200$

9. Define and give the probability for each of the following:

a) Type I error
b) Type II error
c) Power

10. List three things we can do to increase the power of the t-test.

CHAPTER 11

Inference with the F-Distribution

ANOVA: Analysis of Variance.

The previous two chapters dealt with statistical techniques involving one or two groups of observations or samples. Much scientific research requires statistical analysis of more than two samples. For example, what if we were interested in the recovery rates of patients who had psychoanalysis, behavior modification, or gestalt therapy?; or perhaps we would like to compare the running rates of rats in 10 different kinds of mazes. When we have more than two samples and wish to compare their means, we may find a statistical technique called **analysis of variance (ANOVA)** useful. This is a family of statistical techniques specifically designed for experiments with more than two groups of subjects.

ANOVA is the appropriate technique for making inferences about population means of two groups or more and uses various measures of variability to do so, hence the name Analysis of *Variance*. All groups are compared simultaneously and thus, this a very efficient statistical procedure.

You may be wondering why we don't simply compare pairs of samples with several t-tests. This is a good question. Consider an experiment where we gave different doses of an experimental drug to five groups of patients and evaluated the effects on their performance on some motor skills task. We could measure the effect of each dose level by comparing every pair of means using a separate t-test. Remember the formula for the number of combinations of 5 things taken 2 at a time? ${}_5C_2 = 5!/3!2!$ We would need 10 t-tests. There are several problems with this approach. It would be a lot of work, but there is a statistical problem with it, which is more critical. Running multiple t-tests in a single experiment is inappropriate because the probability of an inferential error is high. We will discuss this in more detail later on.

Recall that inference with the normal curve involved computing Z-scores. Inference with the t-distribution required the computation of t-scores. ANOVA is an inferential technique which uses the F-distribution and requires the computation of F-scores.

The F-Distribution

Just as the normal distribution is the random sampling distribution of the Z statistic and the t-distribution is the random sampling distribution of the t statistic, *the F-Distribution is the random sampling distribution of the F statistic.*

In Chapter 10 we saw that when we collect all possible samples from a population, calculate Z and t for each sample mean and place the Z values and the t values in separate relative frequency distributions, the Z values follow a normal curve and the t values follow the t-distribution. We use a similar procedure for setting up the sampling distribution of the F statistic.

Constructing the F-Distribution

The F-Distribution is set up a little differently than what you are used to and, as you learn more about ANOVA, the reasons for this will become clearer. The sampling distribution of the F statistic is constructed in the following way:

Step 1. Randomly select two samples of fixed sizes from a population.

Step 2. Calculate an *unbiased estimate of the variance* (s^2) for each sample.

Step 3. Divide the first variance estimate (s_1^2) by the second variance estimate (s_2^2). *This is the F statistic.*

Step 4. Return the samples to the population.

Step 5. Repeat steps 1-4 until all possible pairs of samples have been drawn.

Step 6. Place the F values in a relative frequency distribution.

The F-distribution has certain important properties that allow us to make inferences about population means.

The F-Distribution is the relative frequency distribution of the F statistic. The F statistic is a ratio between two variance estimates. Since an estimate of the variance cannot be negative (zero variability is as low as it goes), the frequency distribution is limited at zero. The other end of the distribution, however, is not limited and so the distribution is positively skewed.

Notice that the variance estimates are calculated for samples drawn from a single population. On the average, then, we would expect the two estimates to be equal. The expected value of F is approximately 1.00 since we are computing a ratio between two variance estimates each of which estimates the same population variance. In summary:

The F ratio = s_1^2/s_2^2

The F distribution is positively skewed with a limit at zero.

The expected value of F is approximately 1.00.

The F-distribution, like the t-distribution, is a family of curves whose shape depends on degrees of freedom. Here is an example of an F-distribution.

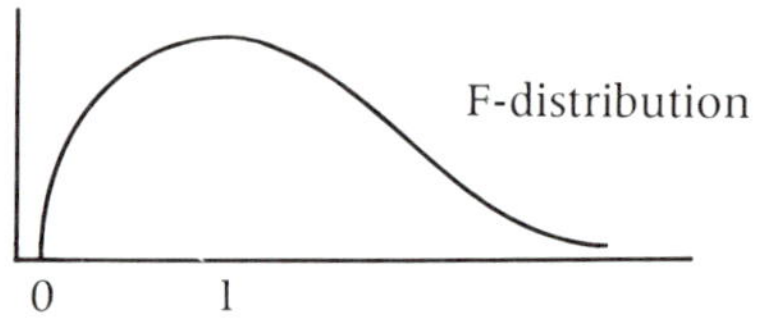

Using the F-Distribution

Recall that ANOVA is used to make inferences about the means of the populations from which samples have been drawn. These different samples have been treated differently by the researcher who wishes to know if the treatment or independent variable (IV) had an effect on the dependent variable (DV). If the IV didn't work, we would expect the samples to be similar and their means to be the same, or differ from each other only by chance. In other words, if the IV has no effect, the samples are simply different samples drawn from the same population and their means will differ due to sampling fluctuation only.

If, however, the treatment has an effect, the samples then come from populations with different means and we would expect the sample means to be quite different.

Are you beginning to see why this analysis is called "Analysis of Variance?" ANOVA examines how much sample means vary from each other compared to what would be expected by chance sampling fluctuation.

Null and Alternate Hypotheses

When we made inferences about means with the normal or the t-distribution, the null hypothesis stated that the means did not differ. The same is true when we make inferences about means using the F-distribution. The null hypothesis is

H_o: $\mu_1 = \mu_2 = \mu_3 = \ldots = \mu_k$
where k is the number of samples in the experiment.

When an analysis compares only two means, we distinguish between directional and non-directional alternate hypotheses. In ANOVA, however, the null hypothesis may be false because two means are different although all the rest are equal, or three means may be different and all the rest the same, for example. ANOVA does not tell us where the difference(s) lies; just that there is a difference between at least two of the means. The alternate hypothesis for ANOVA is

H_a: H_o is false

We do not distinguish between directional and non-directional alternate hypotheses in ANOVA. Once ANOVA has determined that a significant difference exists, we must do further analyses to find out which means are responsible for the outcome of ANOVA. Chapter 13 discusses some of the statistical techniques commonly used after ANOVA.

ANOVA is a family of techniques. We will study two of these in this chapter: **one-way and two-way ANOVA**.

One-Way Analysis of Variance

One-way ANOVA is a statistical procedure that tests for differences between two or more means. With two samples, it is appropriate to compute either t or F.

Comparing two means with a t-test is a special case of one-way ANOVA. In fact, the outcome (not the actual value) of an analysis of variance done on two samples is identical to the outcome of a t-test done on those two samples.

Let's examine the logic of one-way ANOVA. Consider three groups of 10 subjects each. All the subjects were randomly selected from a population and independently assigned to groups. Each group was treated differently. Now imagine, for the moment, that the treatment (IV) had no effect on the DV. We might find the following kinds of results:

```
GROUP A   X XX   XXXXX  X       X
                  X̄a
GROUP B   XXXXX    XX         X  XX
                X̄b
GROUP C   X  X     XXXXX     X  XX
                    X̄c
```

Each X represents the score for a single subject. As you can see, within each group the individual scores vary around the group mean. The means of the three groups are similar but not identical.

Now, let's consider what kind of data we might have if the treatment did have an effect on the groups. We might see something like this.

GROUP A X XX XX XXXXX
$\bar{X}_a$
GROUP B X XXX XXXX X X
$\bar{X}_b$
GROUP C XXX XXXX X XX
$\bar{X}_c$

Within each group, the scores vary around the group mean. The group means are not as similar as when the treatment had no effect. However, we would not expect the means to be identical even if the treatment had no effect. They would vary somewhat because of sampling fluctuation. The question is "how different must the group means be before we conclude that the differences are probably not due to sampling variation alone?" In other words, "how much variation is expected when the groups come from populations with equal means?"

If there is no treatment effect, the populations from which the groups were drawn would have identical means and thus, we may treat the samples as it they came from the same population. Any variation in the sample means would be considered chance variation. If, however, the treatment did have an effect, the samples come from populations with different means and we would expect the variation between the sample means to be quite large.

Partitioning the Variance

Error variance: Variability between subjects that is free of treatment effects.

Looking at the examples given in the previous section, we can identify three kinds of variation. The individual scores *vary around their subgroup means.* The subjects within each group are different and so their scores are different. This is *variation inherent* in the subjects themselves. **Inherent variation** or **error variance** is free of the effects of the treatment. It is the variation we see because of individual differences between our subjects.

> **Total variability:** Variability of all the scores from the combined mean.
>
> **Within group variability:** Variability of scores within groups from the group mean.
>
> **Between group variability:** Variability of group means from combined mean.

The subgroup means are not the same. The *subgroup means vary around the combined mean,* the overall mean of the experiment. These group means are different for two reasons. First, the group means differ partly due to inherent variation of the subjects. Second, they differ due to the effect of the treatment, if indeed it had an effect. The variation between subgroup means is due to *inherent variation plus variation due to treatment.*

Finally, the individual scores vary around the combined mean of the entire distribution. Each score varies from the overall mean. This is the **total variation in the experiment**.

We have three kinds of variation:

1. the total variation as reflected in the difference between the scores and the combined mean;
2. the variation of the scores within each group from their own group mean; and
3. the variation of the group means from the combined mean.

The second kind of variation is due to inherent variability only.

The third kind of variation is due to inherent variation and the effect of the treatment. If the treatment had no effect we would expect the second and third kinds of variation to be equal. In other words, if the treatment had no effect, then these two measures should be more or less the same, given some amount of sampling fluctuation. This is the logic of analysis of variance.

The **total variability** in an experiment may be partitioned thus:

1. the variability of subjects **within groups** and
2. the variability **between groups.**

Any individual score in any group can vary from the combined mean of all the groups $(X - \overline{X}_c)$. Any score can vary from the mean of its own group $(X - \overline{X})$. Each group mean can vary from the combined mean $(\overline{X} - \overline{X}_c)$. If we squared each of these deviation scores and summed the squares, we would have three measures of variability in "sum of squares" form. Analysis of Variance involves computing these various sums of squares (**SS**) and comparing them.

Calculating Sums of Squares Since we have three measures of variability, we compute three separate sums of squares. The measure of total variability (called $\mathbf{SS_{TOT}}$) is composed of two parts: 1. the variability within groups (called $\mathbf{SS_{WG}}$) and 2. the variability between groups (called $\mathbf{SS_{BG}}$). The defining formula which partitions the total variability in an experiment into its two component parts is:

$$SS_{TOT} = SS_{BG} + SS_{WG}$$

$$\Sigma^k\Sigma^n(X_i - \overline{X}_c)^2 = \overset{k}{\Sigma n} \ (\overline{X}_k - \overline{X}_c)^2 + \Sigma^k\Sigma^n(X_i - X_k)^2$$

$\overline{X}_c$ is the combined mean
$\overline{X}_k$ is the group mean
X_i is the raw score

Now, don't let this throw you! Examine the expression to the left of the equal sign. This tells us to subtract the combined mean from each score in the entire experiment, square these differences, and sum them. This is the total variability in the experiment. The first expression to the right of the equal sign tells us to subtract the combined mean from each group mean, square each difference, and sum these.

$$SS_{BG} = (\overline{X}_1 - \overline{X}_c)^2 + (\overline{X}_2 - \overline{X}_c)^2 + \ldots + (\overline{X}_k - \overline{X}_c)^2$$

This is the variability between groups.

The last expression tells us to subtract the group mean from each score in the group, square these differences, sum them, and to do this for all the groups.

$$SS_{WG} = \Sigma x_1^2 + \Sigma x_2^2 + \ldots + \Sigma x_k^2$$

This is the variability within groups.

In the past, we have found that the defining formulas are not usually used computationally. This is true here. There are simplified formulas to be used in the actual computation of the sums of squares.

The computational formula for the total sum of squares is:

$$SS_{TOT} = \Sigma X_{tot}^{\ 2} - \frac{(\Sigma X_{tot})^2}{n_{tot}} \qquad \text{Computational Formula for } SS_{TOT}$$

X_{tot} refers to all the raw scores in the experiment.
n_{tot} refers to the total number of observations in the experiment.

This formula tell us to

1. square all the raw scores in the experiment
2. sum up all the squared raw scores
3. sum up all the raw scores; square the sum, and divide by the total number of observations in the experiment
4. subtract the result of step 3 from that of step 2.

The computational formula for the between group sum of squares is

$$SS_{BG} = \frac{(\Sigma X_1)^2}{n_1} + \frac{(\Sigma X_2)^2}{n^2} + \ldots + \frac{(\Sigma X_k)^2}{n_k} - \frac{(\Sigma X_{tot})^2}{n_{tot}}$$

Computational Formula for SS_{BG}

This formula tells us to

1. sum the raw scores in each group
2. square each group sum and divide each by is group size
3. sum all the values obtained in Step 2
4. sum all the scores in the entire experiment; square this sum and divide by the total number of observations
5. subtract the result of step 4 from that of step 3.

Since the total sum of squares is composed of the SS_{BG} plus the SS_{WG}, we can obtain the sum of squares within by subtraction ($SS_{TOT} - SS_{BG}$). If we wished to calculate the SS_{WG} directly, we would determine the sum of squares within each of our groups using our usual formula for any sum of squares:

$$SS = \Sigma x^2 = \Sigma X^2 - (\Sigma X)^2/n$$

We would compute this value for each group and then sum the resulting values. The formula would look like this

$$SS_{WG} = [\Sigma X_1^2 - (\Sigma X_1)^2/n_1] + [\Sigma X_2^2 - (\Sigma X_2)^2/n_2] + \ldots + [\Sigma X_k^2 - (\Sigma X_k)^2/n_k]$$

Computational Formula for SS_{WG}

In the past, we have divided sums of squares by degrees of freedom to obtain unbiased estimates of the population variance. We will do the same

Mean square: Estimate of the population variance, found by dividing sum of squares by df.

here. Dividing each SS by its associated degrees of freedom produces unbiased estimates of the variance called **mean squares**.

Calculating Mean Squares When we compute the total sum of squares, we effectively ignore the groups and treat all the scores as one large group. We subtract each raw score from the overall or combined mean, square each difference, and sum the squared differences. Because we use the combined mean in our calculation, it is fixed, and we use up one degree of freedom. Therefore, $n_{tot} - 1$ scores are left free to vary. When computing SS_{TOT} then

$$df_{tot} = n_{tot} - 1 \qquad \text{Degrees of Freedom for } SS_{TOT}$$

Computing the sum of squares between groups requires that we subtract the combined mean from each subgroup mean, square each difference, and sum the squared differences. Because we use the combined mean in our computations, it is fixed. We have as many difference values as we have groups, but we have used up one degree of freedom. When computing SS_{BG} then

$$df_{bg} = k - 1 \qquad \text{Degrees of Freedom for } SS_{BG}$$

The sum of squares within groups is found by measuring the amount of variability of the group scores around their own means. We subtract the subgroup mean from all the scores in its group, square the differences, and sum them. We do this for each group in our experiment. Because we use the subgroup means in our calculation, each is fixed, and we use up one degree of freedom for each group mean used. When computing SS_{WG} then

$$df_{wg} = n_1 - 1 + n_2 - 1 + \ldots + n_k - 1$$
$$= n_{tot} - k \qquad \text{Degrees of Freedom for } SS_{WG}$$

In an experiment, the total sum of squares can be partitioned into the sum of squares between, plus the sum of squares within, groups. The total number of degrees of freedom in the experiment can be partitioned into two parts as well: degrees of freedom between plus degrees of freedom within groups. Therefore

$$df_{tot} = df_{bg} + df_{wg}$$
$$n_{tot} - 1 = k - 1 + n_{tot} - k$$

The partitioning of the variability and degrees of freedom in a One-Way ANOVA is illustrated in the following diagram.

One-Way ANOVA

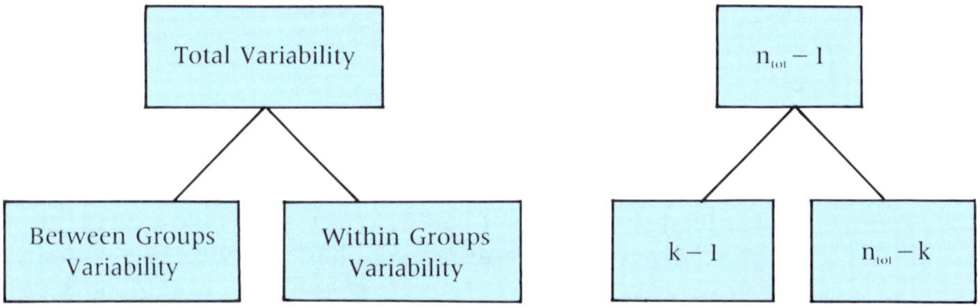

Dividing each sum of squares by its associated degrees of freedom produces unbiased estimates of the population variance: mean squares. If the treatment in our experiment had no effect on the DV, we would expect the estimate of the variance between groups (MS_{BG}) to equal the estimate of the variance within groups (MS_{WG}). This is because the variance estimate between groups reflects inherent variation plus treatment; the variance estimate within groups reflects inherent variation only. If the treatment has no effect, these two measures are estimating the same thing: *inherent variation*. We are interested, then, in these two unbiased estimates: the mean square between, and the mean square within, groups. Each is found by dividing the sum of squares by degrees of freedom.

In summary

$$MS_{BG} = \frac{SS_{BG}}{df} = \frac{SS_{BG}}{k - 1} \qquad \text{Mean Squares One-Way ANOVA}$$

$$MS_{WG} = \frac{SS_{WG}}{df} = \frac{SS_{WG}}{n_{tot} - k}$$

Now we can compare these two variance estimates to determine the effect of the treatment. If the treatment didn't work, each estimates the same population variance. If the treatment did work, then the MS_{BG} would be larger than the MS_{WG}, since it reflects not only inherent variation, but also treatment effects.

Calculating the F-Ratio

The F-ratio is the ratio between the between group variance estimate and the within groups variance estimate:

$$F = MS_{BG}/MS_{WG} \qquad \text{F-Ratio}$$

Clearly, if the treatment had no effect, we would expect this ratio to be 1, on the average. If the treatment did have an effect, we would expect this value to be greater than 1. Table 3, at the back of the book, provides the critical values of F for the 5% and 1% level of significance. This table is read by locating the degrees of freedom associated with the numerator (MS_{BG}) along the top, and locating the degrees of freedom associated with the denominator (MS_{WG}) along the side. As usual, to reject the null hypothesis, the obtained F value must be equal to or larger than the critical value. If the null is rejected, we conclude that the treatment had a significant effect on the DV. In other words, the MS_{BG} is so much greater than the MS_{WG} that we claim that this is not due to chance, but is caused by the treatment variable.

Let's work our way through an example in detail to see how ANOVA is done.

Running a One-Way ANOVA

As a psychology instructor, I am always looking for ways to help students improve their study skills and test performance. Let's assume that I have randomly selected three groups of 15 students each, and I have given special training to two of the groups. In one group, I have concentrated their training on test-taking skills (Group T). Another group has been trained in study skills (Group S). The third group has received no special training. This is the control group (Group C). The trained groups have been instructed to use their new skills when studying and taking weekly tests for a period of one month. At the end of this time all the students take a common test to measure their knowledge about the preceding month's work. Here are the data:

	Group T	Group S	Group C	X_1^2	X_2^2	X_3^2
	64	56	45	4096	3136	2025
	78	76	60	6084	5776	3600
	68	55	65	4624	3025	4225
	57	53	67	3249	2809	4489
	76	76	54	5776	5776	2916
	75	75	53	5625	5625	2809
	73	51	58	5329	2601	3364
	66	62	59	4356	3844	3481
	87	69	49	7569	4761	2401
	89	63	72	7921	3969	5184
	68	52	63	4624	2704	3969
	59	53	69	3481	2809	4761
	72	81	61	5184	6561	3721
	83	90	53	6889	8100	2809
	61	84	58	3721	7056	3364
Sums	1076	996	886	78528	68552	53118
Means	71.73	66.40	59.07			

We have all the raw scores and their squares. Now let's go ahead and run a one-way ANOVA on these data.

Step 1. Compute the Sums of Squares

$$SS_{TOT} = \Sigma X_{tot}^2 - (\Sigma X_{tot})^2/n_{tot}$$

$$\Sigma X_{tot}^2 = 78528 + 68552 + 53118 = 200198$$

$$\Sigma X_{tot} = 1076 + 996 + 886 = 2958$$

$$(\Sigma X_{tot})^2/n_{tot} = 2958^2/45 = 194439$$

$$SS_{TOT} = 200198 - 194439 = 5758.80$$

$$SS_{BG} = \frac{(\Sigma X_1)^2}{n_1} + \frac{(\Sigma X_2)^2}{n_2} + \frac{(\Sigma X_3)^2}{n_3} - \frac{(\Sigma X_{tot})^2}{n_{tot}}$$

$$= \frac{1076^2}{15} + \frac{996^2}{15} + \frac{886^2}{15} - 194439$$

$$= 195653 - 194439 = 1213.33$$

$$SS_{WG} = SS_{TOT} - SS_{BG} = 5758.80 - 1213.33 = 4545.47$$

We can verify the SS_{WG} by computing it directly.

SS within Group T $= \Sigma X_1^2 - (\Sigma X_1)^2/n_1 = 78528 - 1076^2/15 = 1342.93$

SS within Group S $= \Sigma X_2^2 - (\Sigma X_2)^2/n_2 = 2417.6$

SS within Group C $= \Sigma X_3^2 - (\Sigma X_3)^2/n_3 = 784.93$

$SS_{WG} = 4545.47$

Step 2. Compute the Mean Squares
To calculate the F-ratio we use MS_{WG} and MS_{BG}. We have a total of $n_{tot} - 1$ degrees of freedom $(45 - 1 = 44)$. Because we have three groups, the MS_{BG} has 2 degrees of freedom. The within group MS has 42 df $(n_{tot} - k)$.

$$MS_{BG} = SS_{BG}/df = 1213.33/2 = 606.67$$

$$MS_{WG} = SS_{WG}/df = 4545.47/42 = 108.23$$

Step 3. Compute the F-ratio

$$F = MS_{BG}/MS_{WG} = 606.67/108.23 = 5.61$$

Step 4. Compare the obtained F value with the critical value and make a decision.

$$F_{crit} = F_{(2,42)} = 5.15 \text{ at } \alpha = .01$$

Since our obtained value is larger than the critical value, we reject the null hypothesis. We have evidence that there is a significant difference between the means of the populations from which our samples were drawn. We can say that training makes a difference in test performance. The next step would be to find out exactly where that difference lies. This requires that we follow our ANOVA with another statistical technique designed to compare individual means. The more common comparison procedures are discussed in Chapter 13.

The outcome of any ANOVA can be summarized in a table called an **ANOVA Summary Table**. The summary table for the analysis we just did would look like this.

ANOVA Summary Table

Source of Variance	Sum of Squares	df	Mean Squares	F	p
Between groups	1213.33	2	606.60	5.61	<.01
Within groups	4545.47	42	108.23		
Total	5758.80	44			

Summary of One-Way ANOVA

Hypothesis

H_o: $\mu_1 = \mu_2 = \ldots = \mu_k$

H_a: H_o is false

Assumptions

1. Subjects are randomly selected and independently assigned to groups.
2. Population distributions are normal
3. Population variances are homogeneous.

Decision Rules

$df_{bg} = k - 1 \quad df_{wg} = n_{tot} - k$

If $F_{obt} \geq F_{crit}$, reject the H_o

If $F_{obt} < F_{crit}$, do not reject H_o

Formula

$$F = \frac{MS_{BG}}{MS_{WG}}$$

One-way ANOVA is a method for comparing groups receiving different treatments or different levels of *one* independent variable. Another ANOVA technique allows us to simultaneously study the effects of *two* independent variables.

Two-Way Analysis of Variance

Two-way analysis of variance simultaneously measures the effects of two treatment conditions or independent variables. Perhaps we are interested in determining the readability of four computer type styles under fluorescent, versus incandescent light. We could do two experiments: one to study the readability of the four styles under fluorescent light, and a second to study readability under incandescent light. This approach is not very efficient, however. Two-way ANOVA allows us to investigate the two independent variables (light and style) at the same time in a single experiment. This design is particularly important because it provides an analysis of the interaction between the two variables.

The Logic of Two-Way ANOVA

Main effect: In ANOVA, the effect of an independent variable on performance of groups.
Interaction: In ANOVA, the effect of one independent variable on the other.

A two-way ANOVA provides three different F-tests to answer three different questions about the effects of the treatments. Using the previous example, let's consider the questions we can ask. First, does type style make a difference in readability, overall? Second, is there a difference, overall, in readability under incandescent versus fluorescent light? Third, whatever the difference in readability of the four styles, is this difference the same under both light conditions? The first two questions are asking about **main effects**. The third question asks whether or not there is an **interaction** between type style and lighting conditions.

Main Effects Main effect refers to the effect of each independent variable. If type style made a difference in readability, we would say there is a *main effect of style*. If light conditions also made a difference, we would say there is a *main effect of lighting*. The readability scores from our experiment are as follows:

Light	Chicago	Geneva	Monaco	Princeton	Sums	Means
Fluorescent	35	22	20	31	108	27
Incandescent	28	23	8	29	88	22
Sums	63	45	28	60		
Means	31.5	22.5	14.0	30.0		

Let's consider the main effect of type style: Does style make a difference overall? To determine the effect of style we examine the sums (or means) for each style, ignoring light conditions, for the moment. We can see that the readability score is highest for Chicago and lowest for Monaco. Let's graph the main effect of style.

Main effect of style

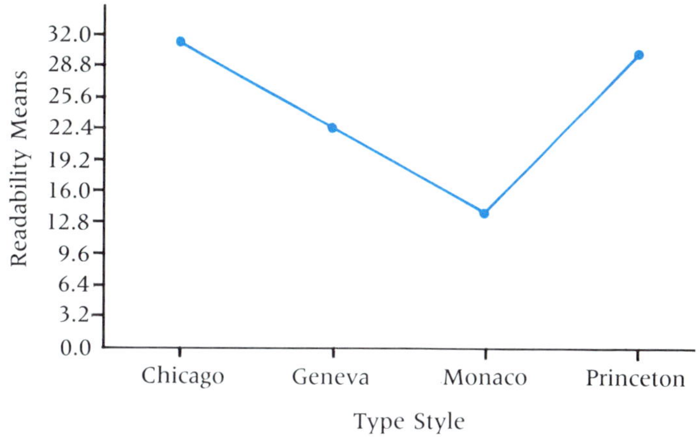

This graph indicates that readability is poorest for Monaco, best for Chicago and Princeton and intermediate for Geneva. If the line was flat, there would be no main effect due to type style.

The second independent variable was light condition. If we graphed the data on readability under the two kinds of lighting, we could examine the main effect of lighting.

Main effect of lighting

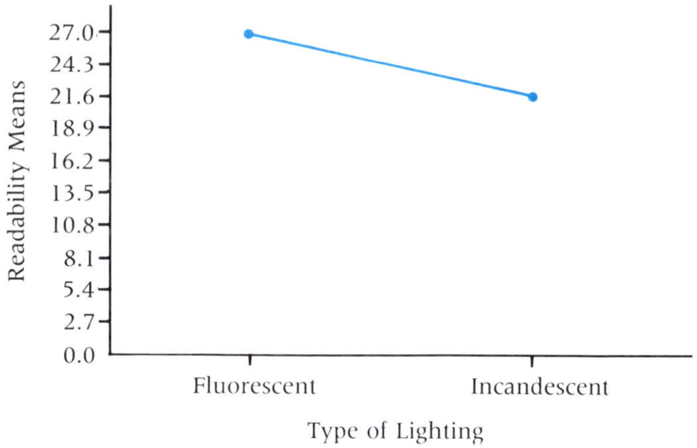

This graph indicates that readability was better under fluorescent lighting conditions. The third question we asked was "Whatever the effect of one of the independent variables, is it the same under the levels of the other IV?" In other words, "Is there an interaction between the two IV's?"

The Interaction An interaction, in our example, means that the effect of type style on readability is different under fluorescent and incandescent lighting. To graph the interaction, we separate the scores under the two lighting conditions.

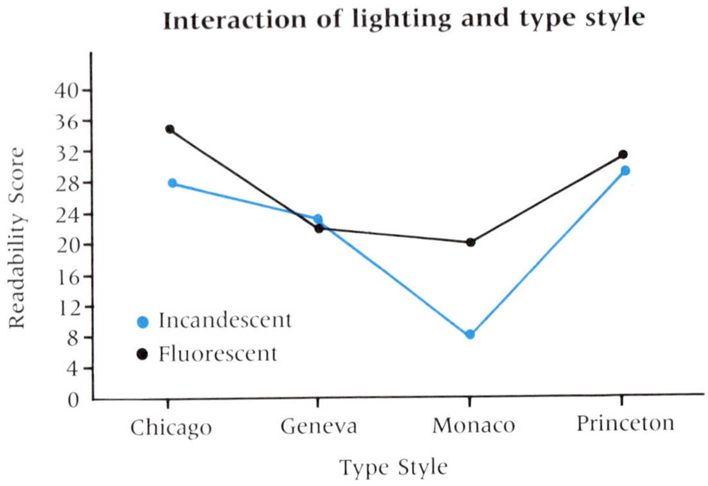

Our graph indicates that the Monaco type style was particularly hard to read under incandescent light. It appears to be the style most affected by the change in lighting conditions. If the effect of lighting was the same for all type styles, the two lines of our graph would be more or less parallel.

The F-test for the interaction measures the degree to which the lines are parallel. If it is significant, we conclude that the lack of parallelism is not due to chance alone. Of course, we would not expect the lines to be exactly parallel; we would expect some deviation. If the deviation is very large, however, we have a significant interaction.

Finding a significant interaction affects how we interpret our data. Consider an example where two teachers are interested in the effectiveness of two instructional methods on student performance. Professor A is very effective when she uses a small-group discussion approach. When she uses a straight lecture approach, however, her students do not do very well. Professor B, on the other hand, is a very effective lecturer and students do much better when he lectures than when he uses a small-group approach. If we obtained overall performance scores from the students under each teaching method, we might find the following situation.

Student performance under two methods

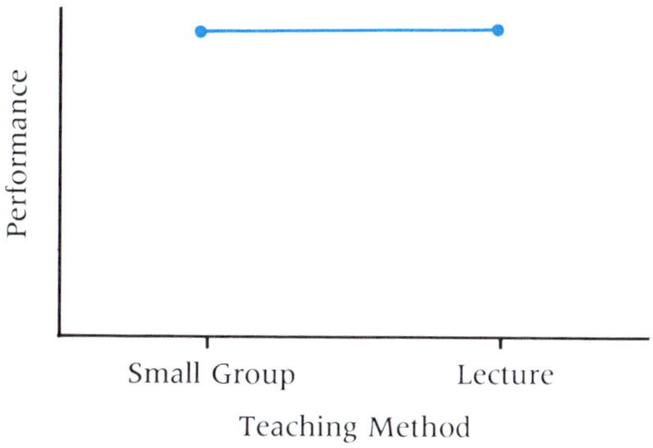

This graph indicates that there is no difference in overall performance under the two teaching methods, that is, there is *no main effect of teaching method*. When we look at the two professors, ignoring method, the data might look like the following.

Student performance with two professors

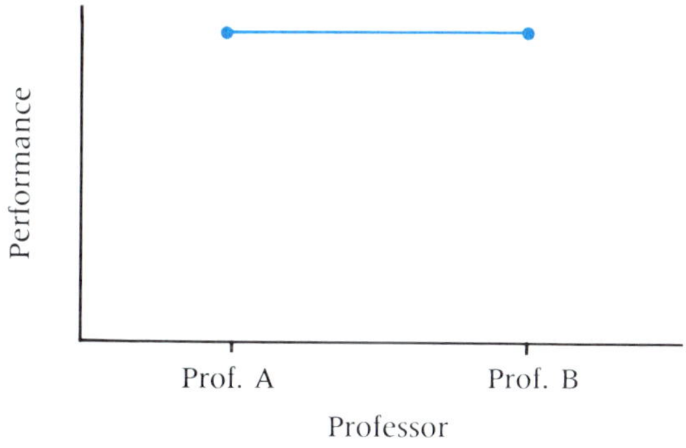

This graph indicates that there is no difference in performance between Professor A's students and Professor B's students, that is, there is *no main effect of teacher*. If our interpretation of the data stopped here, it would be misleading. The lack of main effects does not necessarily mean that there are no differences. We must also examine the interaction between teacher and method. The graph of the interaction would look like the following.

Interaction of teacher with method

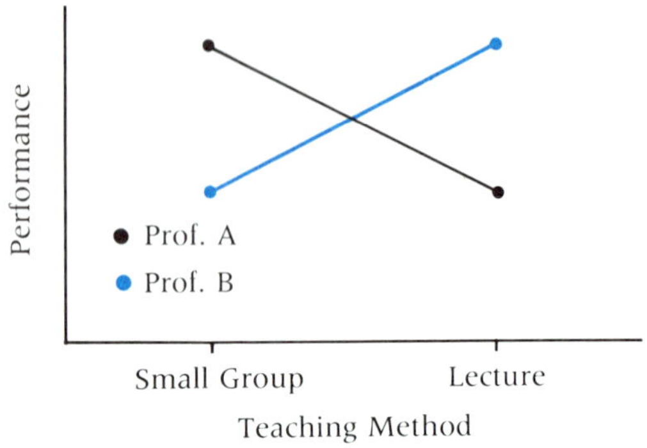

Now we can see what really happened. Professor A was very effective with the small group approach and Professor B was very effective with the lecture approach. The teaching method interacted with the teacher using it. If we had looked only at the main effects we would have concluded, incorrectly, that neither teacher nor method made a difference in student performance. Two-way ANOVA allows us to examine the interaction between the two independent variables. It provides us with important information that we do not have with one-way ANOVA.

Partitioning the Variance

Let's consider an experiment with two independent variables (A and B), each having two levels (1 and 2). This design is called a **2X2 Complete Factorial**. We have two variables or factors, each of which has two levels. In total, we have four groups of subjects. Like one-way ANOVA, we can partition the total variance in the experiment into its component parts. The total sum of squares can be divided into the sum of squares between groups, (often called the treatment sum of squares) and the sum of squares within groups.

What is new with two-way ANOVA is that the *sum of squares between groups can be partitioned into three sums of squares*: $SS_{BG} = SS_A + SS_B + SS_{AXB}$.

1. the sum of squares associated with the first independent variable (SS_A).
2. the sum of squares associated with the second independent variable (SS_B), and
3. the sum of squares associated with the interaction (SS_{AXB}).

Each sum of squares, when divided by its associated degrees of freedom, gives us an unbiased estimate of the population variance. These variance estimates are called mean squares as before.

$MS_A = SS_A/df$
$MS_B = SS_B/df$ Mean Squares for Two-Way ANOVA
$MS_{AXB} = SS_{AXB}/df$

If neither A or B had an effect, then these three mean squares estimate the same population variance as the within-group mean square. With a two-way ANOVA, we have three new tests of significance and three new F-ratios. The F-ratios are determined, as before, by dividing the between-group variance estimates by the within-group variance estimates. In summary

F for the A Main Effect $= MS_A/MS_{WG}$
F for the B Main Effect $= MS_B/MS_{WG}$
F for the Interaction $= MS_{AXB}/MS_{WG}$

F-Ratios for Two-Way ANOVA

Each F-ratio is compared to the critical value in Table 3 to determine its significance.

Let's look at an example to see how we would calculate the sums of squares for our three tests of significance.

Calculating the Sums of Squares

Below is an example of a 2x2 Factorial Design layout. Thirty-two subjects are randomly selected and independently assigned to four groups, with eight subjects in each.

	GROUPS			
	A_1		A_2	
Subject #	B_1	B_2	B_1	B_2
1				
2				
3				
4				
5				
6				
7				
8				
Sums	A_1B_1	A_1B_2	A_2B_1	A_2B_2

Step 1. Calculate the total sum of squares. This is done in the same way as before. Recall the formula for SS_{TOT}

$$SS_{TOT} = \Sigma X_{tot}^2 - (\Sigma X_{tot})^2/n_{tot}$$

Once again, we square all the raw scores, add them up and subtract from this value the square of the sum divided by the total number of scores in the experiment.

Step 2. Calculate the between-groups and the within-groups sum of squares. We use the same procedure as we did with one-way ANOVA. We'll modify our earlier formula for the between sum of squares to avoid confusion.

The between-group sum of squares formula for a general two-way ANOVA is

$$SS_{BG} = \frac{(\Sigma A_1B_1)^2 + (\Sigma A_1B_2)^2 + \ldots + (\Sigma A_aB_b)^2}{n} - \frac{(\Sigma X_{tot})^2}{n_{tot}}$$

where n is the number of subjects in each group; a is the number of levels of the A independent variable; and b is the number of levels of the B independent variable.

And so, ΣA_aB_b is the sum of the scores in the last group. For the design presented above, $\Sigma A_aB_b = \Sigma A_2B_2$, the fourth group in the experiment. Obviously, for this design ab is the total number of groups in the experiment, which we often call k.

The within-group sum of squares is easily found by subtraction.

$$SS_{WG} = SS_{TOT} - SS_{BG}$$

Step 3. Calculate the sums of squares for the main effects and the interaction. Recall that we have three between-group tests of significance because we partitioned the between-group sum of squares into three parts: the two main effects and the interaction. We must determine the sum of squares for the A main effect, for the B main effect, and for the AxB interaction. The general formula for computing the sum of squares for the A main effect is

$$SS_A = \frac{(\Sigma A_1)^2 + (\Sigma A_2)^2 + \ldots + (\Sigma A_a)^2}{bn} - \frac{(\Sigma X_{tot})^2}{n_{tot}}$$

SS for the A Main Effect: Two-Way ANOVA

where bn is the total number of subjects in each level of the A independent variable.

For the A main effect, in our 2x2 design, we must sum over the two levels of B for A_1 and over the two levels of B for A_2, square each sum, and divide by the number of subjects contributing to that sum, i.e., bn. We ignore the B independent variable and look only at the difference between the levels of A. We then subtract the same term: the square of the sum of all the scores in the experiment divided by the total number of scores. Our formula is

$$SS_A = \frac{(\Sigma A_1)^2 + (\Sigma A_2)^2}{bn} - \frac{(\Sigma X_{tot})^2}{n_{tot}}$$

To calculate the B main-effect sum of squares, the general formula is

$$SS_B = \frac{(\Sigma B_1)^2 + (\Sigma B_2)^2 + \ldots + (\Sigma B_b)^2}{an} - \frac{(\Sigma X_{tot})^2}{n_{tot}}$$

SS for the B Main Effect: Two-Way ANOVA

where an is the total number of subjects in each level of B.

For our design, this formula is simplified to

$$SS_B = \frac{(\Sigma B_1)^2 + (\Sigma B_2)^2}{an} - \frac{(\Sigma X_{tot})^2}{n_{tot}}$$

Here we ignore the A IV and look only at the difference between the two levels of B.

Remember that the between-groups sum of squares is composed of the sum of squares for A, the sum of squares for B and the sum of squares for the AxB interaction. The interaction sum of squares, then, can be found by subtraction.

$$SS_{AXB} = SS_{BG} - SS_A - SS_B$$

SS for the Interaction Two-Way ANOVA

Calculating the Mean Squares

Each sum of squares is divided by its degrees of freedom to provide mean squares. Recall that the total degrees of freedom in the experiment is $n_{tot} - 1$. The within-group sum of squares has $n_{tot} - k$ degrees of freedom, and the between-group sum of squares has $k - 1$ degrees of freedom. Recall that k for a two-way ANOVA is the product of a and b (i.e., the number of levels of A and the number of levels of B).

Since the between-group sum of squares is partitioned into three sums of squares, we might expect the degrees of freedom between groups to be partitioned as well; and this is so. The degrees of freedom for the A main effect are found by subtracting 1 from the number of levels of A. If there are two levels of A, we have 1 df for the A main effect. The same is true for the main effect of B. The degrees of freedom are the number of levels of B minus 1. The remaining degrees of freedom between groups are taken by the interaction. The interaction, then, has $k - 1 - df_a - df_b$. The degrees of freedom for the interaction can also be obtained by $(df_a)(df_b)$. A 2x2 experiment with four groups would have 3df between groups: 1df for the A main effect, 1df for the B main effect and 1df for the interaction. In summary:

$$df_{bg} = df_a + df_b + df_{axb}$$

The mean squares are now easy to obtain.

Between-Group Means Squares for Two-Way ANOVA

$$MS_A = \frac{SS_A}{df_a}$$ df is $a - 1$, where a is the number of levels of A

$$MS_B = \frac{SS_B}{df_b}$$ df is $b - 1$, where b is the number of levels of B

$$MS_{AXB} = \frac{SS_{AXB}}{df_{axb}}$$ df is $(df_{bg} - df_a - df_b)$ or $(df_a)(df_b)$

Partitioning of the variability and degrees of freedom for a two-way ANOVA is illustrated in the following diagram.

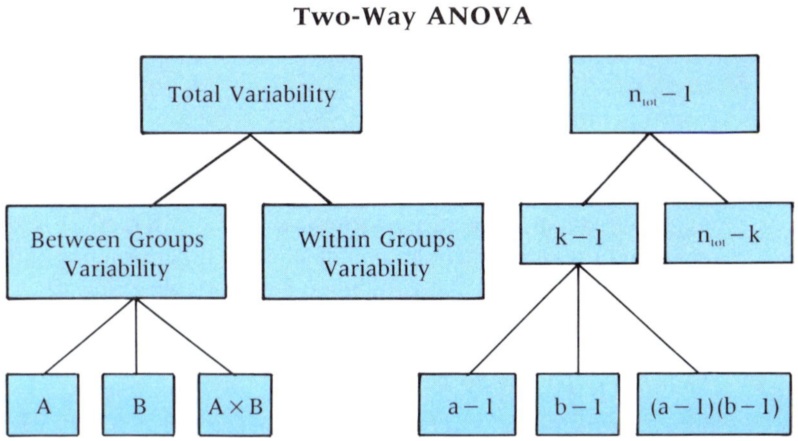

Calculating the F-Ratios

Now we can test the three mean squares for significance.

$$F_A = MS_A/MS_{WG}$$
$$F_B = MS_B/MS_{WG} \qquad \text{F-Ratios for Two-Way ANOVA}$$
$$F_{AXB} = MS_{AXB}/MS_{WG}$$

All three F-ratios use the MS within groups in the denominator. Each is tested for significance by locating the appropriate degrees of freedom in Table 3. The between-group, or treatment-mean square itself, may be tested for significance with the within-group mean square in the denominator. Lack of significance of the between-group mean square does not preclude a significant result in the other tests, however. Often, we don't test the between-group mean square in a two-way ANOVA, since our interest is in the main effects and the interaction rather than the overall treatment-mean square.

Interpreting the F-ratios

As in one-way ANOVA, we would expect the F ratios to be close to 1, on the average, if the independent variables had no effect on performance. With a two-way ANOVA, however, we have three questions to ask.

1. Did our first independent variable have an overall effect?
2. Did our second independent variable have an overall effect?
3. Was the effect of the first independent variable the same under each level of the second independent variable?

If the F ratio for the A main effect was significant, then we know that A made a difference in performance. If the B main effect was also significant, we know that B affected performance. If the interaction is significant, we know that the A independent variable affected performance differently under the different levels of the B independent variable.

Running a Two-Way ANOVA

Let's do a two-way ANOVA. Once again we'll use a 2x2 factorial design. A researcher decides to investigate the effect of practice and corrective feedback on problem-solving performance in elementary school children. She randomly selects 32 children and divides them into four groups of 8 each. Two of the groups are given 25 practice problems to work on and the other two groups are given 5 practice problems. In addition, she provides corrective feedback to one of the high-practice groups and to one low-practice group. The remaining two groups receive no feedback. Here are her data.

	No Feedback		Corrective Feedback	
	Hi Prac.	Low Prac.	Hi Prac.	Low Prac.
	B_1	B_2	B_1	B_2
	6	6	12	9
	9	5	9	8
	6	7	9	8
	9	4	8	9
	8	6	7	7
	6	2	9	9
	8	4	8	5
	4	7	6	6
Sums	56	41	68	61
Means	7.00	5.13	8.50	7.63

$\Sigma X_{tot} = 56 + 41 + 68 + 61 = 226$

Step 1. Compute the total sum of squares.

$$SS_{TOT} = \Sigma X_{tot}^2 - (\Sigma X_{tot})^2/n_{tot}$$

$$SS_{TOT} = 6^2 + 9^2 + 6^2 + \ldots + 6^2 - 226^2/32$$

$$= 1726 - 1596.13 = 129.88$$

Step 2. Compute the between group sum of squares.

$$SS_{BG} = \frac{(\Sigma A_1B_1)^2 + (\Sigma A_1B_2)^2 + \ldots + (\Sigma A_aB_b)^2}{n} - \frac{(\Sigma X_{tot})^2}{n_{tot}}$$

$$SS_{BG} = \frac{56^2 + 41^2 + 68^2 + 61^2}{8} - \frac{226^2}{32}$$

$$= 1645.25 - 1596.12 = 49.13$$

Step 3. Compute the within group sum of squares.

$$SS_{WG} = SS_{TOT} - SS_{BG}$$

$$= 129.88 - 49.13 = 80.75$$

Step 4. Compute the A, B and AXB sums of squares.

$$SS_A = \frac{(\Sigma A_1)^2 + (\Sigma A_2)^2}{bn} - \frac{(\Sigma X_{tot})^2}{n_{tot}}$$

$$= \frac{97^2 + 129^2}{16} - \frac{226^2}{32}$$

$$= 32$$

$$SS_B = \frac{(\Sigma B_1)^2 + (\Sigma B_2)^2}{an} - \frac{(\Sigma X_{tot})^2}{n_{tot}}$$

$$= \frac{124^2 + 102^2}{16} - \frac{226^2}{32}$$

$$= 15.13$$

$$SS_{AXB} = SS_{BG} - SS_A - SS_B$$

$$= 49.13 - 32 - 15.13 = 2$$

Step 5. Compute the mean squares.

$$MS_A = SS_A/df = 32/1 = 32$$

$$MS_B = SS_B/df$$

$$= 15.13/1 = 15.13$$

$$MS_{AXB} = SS_{AXB}/df$$

$$= 2/1 = 2$$

$$MS_{WG} = SS_{WG}/df$$

$$= 80.75/28 = 2.88$$

Step 6. Compute the F-ratios.

$$F_A = MS_A/MS_{WG} = 32/2.88 = 11.10$$

$$F_B = MS_B/MS_{WG} = 15.13/2.88 = 5.25$$

$$F_{AXB} = MS_{AXB}/MS_{WG} = 2/2.88 = 0.69$$

Step 7. Complete the ANOVA Table.

The outcome of any analysis of variance is presented in a summary table.

ANOVA Summary Table

Source	SS	df	MS	F	p
Between					
A	32.00	1	32.00	11.10	< .01
B	15.13	1	15.13	5.25	< .05
AXB	2.00	1	2.00	0.69	> .05
Within	80.75	28	2.88		
Total	129.88	31			

What does all this mean? Our researcher has two significant main effects, one at $\alpha = .05$ and one at $\alpha = .01$, and no significant interaction. The A main effect tells her that providing feedback made a significant difference in performance. The B main effect tells her that practice also made a difference. The lack of significance of the interaction means that the two variables, practice and feedback, did not interact. The effect of feedback was similar under high and low practice conditions. Let's graph both main effects and the interaction to verify this.

Main effect of feedback

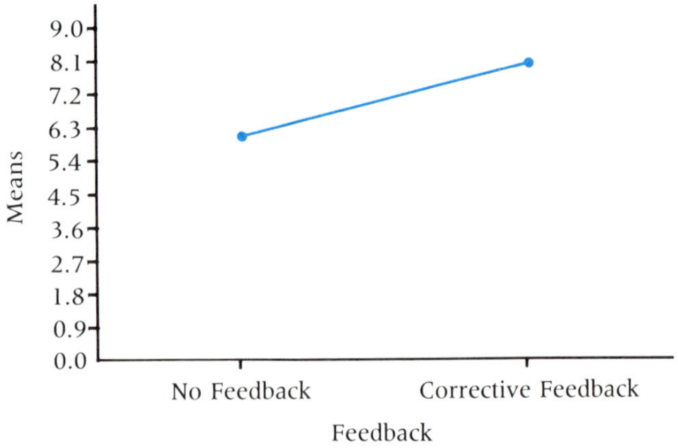

We can see that the groups given corrective feedback performed better than those not given feedback. Since the F-ratio was significant, we can claim that feedback improved performance. Let's look at the main effect of practice.

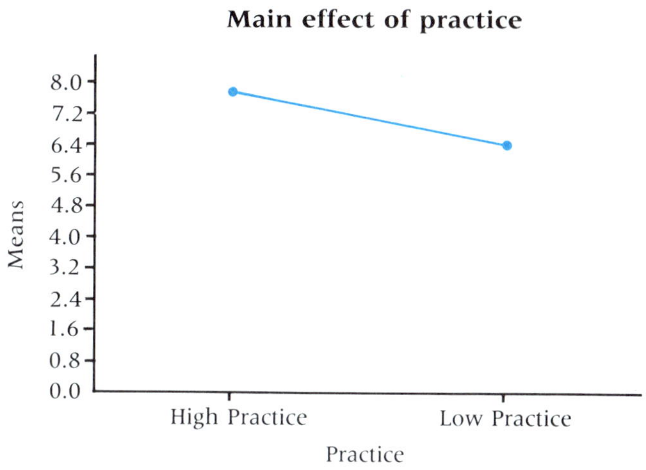

We can see that children given many practice problems did better than those given only a few; this is supported by a significant F-ratio.

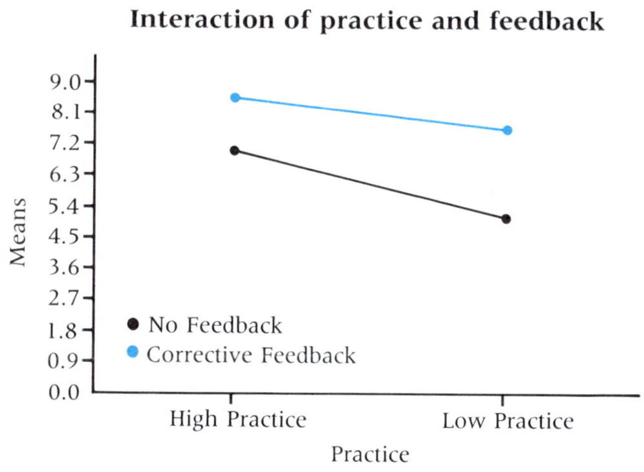

Recall that the test of significance of the interaction involves testing whether the two lines depart significantly from parallelism. The interaction is not significant; the two lines do not depart from parallel more than expected by chance. They don't look exactly parallel but, nevertheless, they are not divergent enough for significance. There is no evidence that the effect of feedback was different under the two practice conditions.

You may have noticed that all the examples of two-way ANOVA have used groups of equal sizes. This is not necessary; however, the calculations for a design with unequal group sizes is beyond the scope of this book.

Summary of Two-Way ANOVA

Hypotheses

H_o: No main effects and no interaction.

H_a: H_o is false.

Assumptions

1. Subjects are randomly selected and independently assigned to groups.
2. Population distributions are normal.
3. Population variances are homogeneous.

Decision Rules

$df_a = a - 1 \qquad df_b = b - 1 \qquad df_{axb} = (a - 1)(b - 1)$

If $F_{obt} \geq F_{crit}$, reject the H_o

If $F_{obt} < F_{crit}$, do not reject H_o

Formulas

$$F_A = \frac{MS_A}{MS_{WG}}$$

$$F_B = \frac{MS_B}{MS_{WG}}$$

$$F_{AXB} = \frac{MS_{AXB}}{MS_{WG}}$$

Assumptions Underlying Inference with the F-Distribution

The analysis of variance techniques we have studied assume that the populations, from which the samples were drawn, are normally distributed. The central limit theorem assures us that the assumption of normality won't be seriously violated as long as our samples are reasonably large and of the same size.

ANOVA also assumes that the populations from which the samples are drawn have equal variances. The F-test is quite robust, however, in that it is relatively insensitive to heterogeneity of variances as long as subjects have been randomly selected and independently assigned to groups.

Inferential Error

Experiment-wise error rate: Probability of Type I error overall in an experiment.

One of the important concerns in research is that of inferential error. You may recall the discussion, in Chapter 9, of Type I and Type II errors in statistical inference. A Type I error occurs, when a true null hypothesis is rejected. The Type I error rate is set by alpha. Earlier in this chapter, I mentioned that one of the problems with using multiple t-tests was the increased probability of making an error. When many t-tests are run within a single experiment, the probability of rejecting at least one true null hypothesis is higher than alpha. The advantage of ANOVA is that the probability of a Type I error is maintained at alpha, over the entire experiment. This is often referred to as the **experiment-wise error rate** (i.e., the probability of making at least one Type I error). In ANOVA, the experiment-wise error rate is $\leq \alpha$.

This chapter has dealt with statistical techniques for research involving more than two groups of different subjects. The following chapter describes the use of analysis of variance in research where subjects serve in more than one group. We have called these "dependent groups designs." They are also known as **repeated measures** designs.

SUMMARY OF FORMULAS

One-Way ANOVA: Computational Formulas

Sums of Squares

Total $$SS_{TOT} = \Sigma X_{tot}^2 - \frac{(\Sigma X_{tot})^2}{n_{tot}}$$

Between Groups $$SS_{BG} = \frac{(\Sigma X_1)^2}{n_1} + \frac{(\Sigma X_2)^2}{n_2} + \ldots + \frac{(\Sigma X_k)^2}{n_k} - \frac{(\Sigma X_{tot})^2}{n_{tot}}$$

Within Groups $$SS_{WG} = SS_{TOT} - SS_{BG}$$

Mean Squares

Between Groups $$MS_{BG} = \frac{SS_{BG}}{k - 1}$$

Within Groups $$MS_{WG} = \frac{SS_{WG}}{n_{tot} - k}$$

F Ratio $$F = MS_{BG}/MS_{WG}$$

Two-Way ANOVA: Computational Formulas

Sums of Squares

Total $$SS_{TOT} = \Sigma X_{tot}^2 - \frac{(\Sigma X_{tot})^2}{n_{tot}}$$

Between Groups $$SS_{BG} = \frac{(\Sigma A_1B_1)^2 + (\Sigma A_1B_2)^2 + \ldots + (\Sigma A_aB_b)^2}{n} - \frac{(\Sigma X_{tot})^2}{n_{tot}}$$

A $$SS_A = \frac{(\Sigma A_1)^2 + (\Sigma A_2)^2 + \ldots + (\Sigma A_a)^2}{bn} - \frac{(\Sigma X_{tot})^2}{n_{tot}}$$

B $$SS_B = \frac{(\Sigma B_1)^2 + (\Sigma B_2)^2 + \ldots + (\Sigma B_b)^2}{an} - \frac{(\Sigma X_{tot})^2}{n_{tot}}$$

AXB $$SS_{AXB} = SS_{BG} - SS_A - SS_B$$

Within Groups $$SS_{WG} = SS_{TOT} - SS_{BG}$$

Mean Squares

A $$MS_A = \frac{SS_A}{a - 1}$$

B $$MS_B = \frac{SS_B}{b - 1}$$

AXB $$MS_{AXB} = \frac{SS_{AXB}}{(a - 1)(b - 1)}$$

WG $$MS_{WG} = \frac{SS_{WG}}{n_{tot} - k}$$

F Ratios

$$MS_A / MS_{WG}$$

$$MS_B / MS_{WG}$$

$$MS_{AxB} / MS_{WG}$$

EXERCISES

1. The data below reflect the outcome of a one-way ANOVA.

$SS_{BG} = 141.3$
$SS_{WG} = 1278$
$k = 4$
$n_1 = n_2 = n_3 = n_4 = 25$

a) Determine the MS_{BG}.
b) Determine the MS_{WG}.
c) Determine F.
d) What is the critical F value at $\alpha = .05$?
e) What is your decision?

2. Complete the one-way analysis of variance for the following data.

$SS_{TOT} = 540.30$
$SS_{BG} = 433.20$
$\alpha = .01$
$n_1 = 5 \quad n_2 = 6 \quad n_3 = 4$

3. An investigator measures the running speed through a maze of two groups of cockroaches. One group had been pre-trained in the maze and the other group was a control group. Run a one-way ANOVA at $\alpha = .05$ on the data. (Note: With two groups a t-test or an ANOVA may be run).

Pre-trained roach speeds in seconds: 7, 9, 12, 10, 10, 9, 8, 8, 8, 4
Control roach speeds in seconds: 9, 11, 10, 10, 13, 14, 12, 12, 13, 14

4. Sketch a rough graph for each of the following outcomes of a two-way ANOVA:

a) One main effect and no interaction.
b) Two main effects and no interaction.
c) One main effect and an interaction.
d) Two main effects and an interaction.

5. The following data were obtained from a two-way ANOVA. Sketch each main effect and the interaction.

Mean Score

	B_1	B_2
A_1	22	3
A_2	14	8

6. A social psychologist was interested in the effects of televised violence on the aggressive behavior of children. She was also interested in the effects of the presence of an authority figure on their behavior. She decided to investigate the two variables simultaneously. She randomly selected four groups of 8 children each. Two of the groups were exposed to a violent cartoon show (V groups) and the other two groups were shown a neutral cartoon show (N groups). All the children were then allowed to play in a common area. Their behavior was monitored by a "blind" judge who counted the number of aggressive acts performed by each child. For one V group and one N group, an adult authority figure was present during the play period. For the other two groups no authority figure was present. Run a two-way ANOVA on the data at $\alpha = .05$.

Violent Groups		Neutral Groups	
Authority	No Authority	Authority	No Authority
2	6	0	0
1	3	1	2
3	7	2	3
2	5	0	1
3	4	1	1
1	6	2	3
2	4	1	1
1	5	3	4

CHAPTER 12

Analysis of Variance with Repeated Measures

Researchers in psychology and related disciplines tend to study organisms that are quite variable in their behavior even under identical experimental conditions. People, for example, are generally more variable than are chemical compounds. People bring their different backgrounds and experiences into the experimental situation. In Chapter 10, we discussed the use of dependent groups designs to reduce the effects of individual differences as a source of variability.

Recall that ANOVA is a family of analyses; some of these are appropriate when the same subjects serve under more than one experimental condition. Because these designs measure the same subjects' behaviour under different conditions, these are called **repeated measures** designs. We will look at two repeated measures designs for analysis of variance.

One-Way ANOVA with Repeated Measures

In one-way ANOVA with repeated measures, all subjects serve under two or more levels (called treatments) of one independent variable. The order of treatment may not be the same for all subjects but, rather, is counterbalanced or randomized in some way. Because each subject is tested under all treatments, differences in performance between the subjects should not reflect treatment differences, but should represent individual differences. The repeated measures design allows us to remove the variability due to individual differences from the estimate of experimental error and, thereby, is a more powerful test of the treatment effect. We assume that the error variance based on repeated measures of the same subject, when the effects of the treatment are removed, will be less than the error variance based on different subjects under the same treatment.

Null and Alternate Hypotheses

The null and alternate hypotheses are the same as those for any one-way ANOVA. The null assumes no difference between population means and the alternate specifies that the null is false. As with other ANOVAs, specific differences between treatment means must be determined by a multiple comparison procedure.

Partitioning the Variance

The total variability in a one-way repeated measures experiment can be partitioned into two parts. The variability between subjects reflects individual differences. The variability within subjects reflects variation due to differences in treatments and to uncontrolled variation when the same subject is tested repeatedly.

The second component, variability within subjects, can be partitioned into 1) variability between treatments and 2) error variability (called subject by treatment or SXT). As you might expect, degrees of freedom can be partitioned similarly. The following diagram illustrates the partitioning of variability and degrees of freedom for a one-way ANOVA with repeated measures.

One-way ANOVA–repeated measures

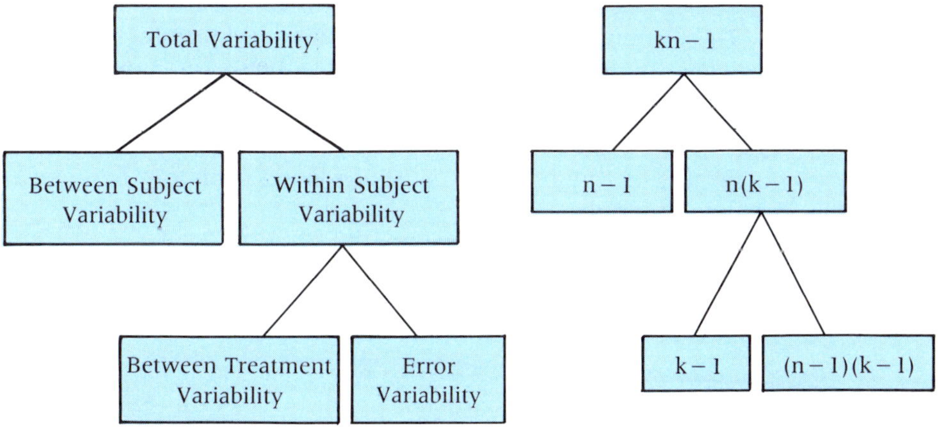

where K is the number of treatment groups and n is the number of subjects in the experiment.

Calculating the Sums of Squares

Below is an example of a one-way repeated measures design with three levels of the independent variable and five subjects tested under each level.

	Treatments			
Subjects	1	2	3	*Sums*
1				S_1
2				S_2
3				S_3
4				S_4
5				S_5
Sums	T_1	T_2	T_3	

Step 1. Calculate the total sum of squares.
This is done in the usual way. We square all the scores, sum them, and subtract from this value the square of the sum of all the scores divided by the total number of scores (kn) in the experiment.

$$SS_{TOT} = \Sigma X_{tot}{}^2 - \frac{(\Sigma X_{tot})^2}{kn}$$

Step 2. Calculate the between subject sum of squares. For simplicity we'll call this the subject sum of squares. The general formula is

$$SS_S = \frac{(\Sigma S_1)^2 + (\Sigma S_2)^2 + \ldots + (\Sigma S_n)^2}{k} - \frac{(\Sigma X_{tot})^2}{kn}$$ Sum of Squares Subjects

This formula tells us to 1) square each subject's sum, 2) sum the squares, 3) divide this sum by the number of treatments (k), and 4) subtract from the value obtained in step 3, the square of the sum of all the scores divided by the total number of scores.

Step 3. Calculate the within subject sum of squares. This can be done with the following formula:

$$SS_{WS} = \Sigma X_{tot}{}^2 - \frac{(\Sigma S_1)^2 + (\Sigma S_2)^2 + \ldots + (\Sigma S_n)^2}{k}$$

You can see that both parts of this formula have been calculated previously in determining the total and the subject sum of squares. Since the total sum of squares is composed of the within subject and the subject sum of squares, the within subject sum of squares can be easily obtained by subtraction.

$$SS_{WS} = SS_{TOT} - SS_S \qquad \text{Sum of Squares Within Subjects}$$

Step 4. Partition the within subject sum of squares into the between treatment or simply, treatment, sum of squares and the error or SXT sum of squares. The formula for the between treatment sum of squares is

$$SS_T = \frac{(\Sigma T_1)^2 + (\Sigma T_2)^2 + \ldots + (\Sigma T_k)^2}{n} - \frac{(\Sigma X_{tot})^2}{kn} \qquad \text{Sum of Squares Treatment}$$

This formula tells us to 1) square each treatment sum and sum the squares, 2) divide the value from 1) by the number of subjects (n), and 3) subtract from this sum the square of the sum of all the scores divided by the total number of scores. This is the source of variability we are interested in since it reflects the effect of the experimental treatment.

The error or SXT sum of squares is easily found by subtracting the treatment sum of squares from the within subject sum of squares.

$$SS_{SXT} = SS_{WS} - SS_T \qquad \text{Sum of Squares Subject by Treatment}$$

This source of variability will be used in the denominator of the F-ratio to test for the significance of the treatment effect.

Calculating the Mean Squares and the F-ratio

The two mean squares used in the F-ratio are obtained as always by dividing the sum of squares by the appropriate degrees of freedom.

The between treatment mean square is

$$MS_T = \frac{SS_T}{k - 1} \qquad \text{Mean Square Treatment}$$

The SXT or error mean square is

$$MS_{SXT} = \frac{SS_{SXT}}{(n - 1)(k - 1)} \qquad \text{Mean Square Error}$$

Notice that the degrees of freedom for the SXT sum of squares are the product of the degrees of freedom for the subject sum of squares (i.e., $n - 1$) and the treatment sum of squares (i.e., $k - 1$).

The F-ratio to test the significance of the experimental treatment is found in the usual way.

$$F = \frac{MS_T}{MS_{SXT}} \qquad \text{F-Ratio}$$

The obtained F value is evaluated for significance against the critical value of F from Table 3 with $k - 1$ df in the numerator and $(n - 1)(k - 1)$ df in the denominator.

Running a One-Way ANOVA with Repeated Measures

The following data are the scores obtained by ten subjects under each of three treatment conditions. Subjects were tested for problem solving performance, alone, with one other person present, or with ten other people present. The researcher was interested in the effect of the presence of others (i.e., the treatment) on performance. The order of the three treatments was randomized for each subject.

	Treatment			
Subject	Alone 1	One Other Present 2	Ten Others Present 3	Sum
1	90	90	70	250
2	95	85	85	265
3	95	70	65	230
4	85	90	65	240
5	95	75	30	200
6	85	85	60	230
7	85	70	75	230
8	80	80	70	230
9	85	80	80	245
10	95	90	75	260
Sums	890	815	675	2380

Step 1. Compute the total sum of squares.

$$SS_{TOT} = \Sigma X_{tot}^2 - \frac{(\Sigma X_{tot})^2}{kn}$$

$$= 90^2 + 95^2 + \ldots + 75^2 - 2380^2/30$$

$$= 194100 - 188813.33$$

$$= 5286.67$$

Step 2. Compute the subject sum of squares.

$$SS_S = \frac{(\Sigma S_1)^2 + (\Sigma S_2)^2 + \ldots + (\Sigma S_{10})^2}{k} - \frac{(\Sigma X_{tot})^2}{kn}$$

$$= \frac{250^2 + 265^2 + \ldots + 260^2}{3} - \frac{2380^2}{30}$$

$$= 189850 - 188813.33$$

$$= 1036.67$$

Step 3. Compute the within subject sum of squares.

$$SS_{WS} = SS_{TOT} - SS_S = 5286.67 - 1036.67 = 4250$$

Step 4. Compute the treatment and subject by treatment sum of squares.

$$SS_T = \frac{(\Sigma T_1)^2 + (\Sigma T_2)^2 + \ldots + (\Sigma T_3)^2}{n} - \frac{(\Sigma X_{tot})^2}{kn}$$

$$= \frac{890^2 + 815^2 + 675^2}{10} - \frac{(2380)^2}{30}$$

$$= 191195 - 188813.33 = 2381.67$$

$$SS_{SXT} = SS_{WS} - SS_T = 4250 - 2381.67 = 1868.33$$

Step 5. Compute the mean squares and F-ratio.

$$MS_T = \frac{SS_T}{k - 1}$$

$$= 23.67/2 = 1190.83$$

$$MS_{SXT} = \frac{SS_{SXT}}{(n - 1)(k - 1)}$$

$$= 1868.33/18 = 103.80$$

$$F = MS_T/MS_{SXT} = 11.47$$

With 2 and 18 degrees of freedom the obtained F value is larger than the critical value at the .01 level of significance, and so we reject the null hypothesis. We have statistical evidence that the presence of others affects problem solving performance. We would report our analysis in an ANOVA summary table.

ANOVA Summary Table

Source	SS	df	MS	F	p
Between Subjects	1036.67	9			
Within Subjects	4250.00	20			
Treatment	2381.67	2	1190.83	11.47	<.01
SXT	1868.33	18	103.80		
Total	5286.67	29			

Summary of One-Way ANOVA with Repeated Measures

Hypotheses

H_o: $\mu_1 = \mu_2 = \ldots = \mu_k$

H_a: H_o is false

Assumptions

1. Subjects are randomly selected.
2. Population distributions are normal.
3. Population variances are homogeneous.
4. Population covariances are equal.

Decision Rules

$df_t = k - 1 \quad df_{sxt} = (n - 1)(k - 1)$

If $F_{obt} \geq F_{crit}$, reject the H_o

If $F_{obt} < F_{crit}$, do not reject H_o

Formula

$$F = \frac{MS_T}{MS_{SXT}}$$

Two-Way ANOVA with Repeated Measures on One Factor

In Chapter 11 we discussed two-way ANOVA as a useful design for simultaneously analyzing the effects of two independent variables with three tests of significance: two tests for main effects due to each independent variable, and the test for the interaction between the independent variables.

Two-way ANOVA can be used to analyze experiments where repeated measures are taken on one or both independent variables. We will concern ourselves with the case where one factor has repeated measures and the other does not. This is quite a common design in several areas of research. We may be interested in the effect of training on performance of groups of subjects who differ with respect to job experience. Perhaps we are investigating the effects of four drug dosages on visual acuity performance of three groups of subjects, each group having received different instructions about the task. If we randomly assigned subjects to instruction groups and tested each under all four dosage levels, we would have a 4x3 factorial design with repeated measures on one factor (drug dosage).

A 2x3 experiment with repeated measures on factor B may be represented as

	B_1	B_2	B_3
A_1			
A_2			

This repeated measures design reduces error variance for tests of the within subjects factors. Differences in performance found between the different levels of factor B under each level of A, for example, cannot be due to differences between subjects, since the same subjects serve under all levels of B. In other words, variability due to individual differences is removed from the analysis of the main effect of factor B. This is also true of the interaction effect. Statisticians say that the B treatment effect and the AXB interaction is unconfounded by between-subject differences. For this reason the test of significance for factor B and the AXB interaction is more sensitive than the test of significance for the A main effect. This design is particularly suitable for researchers who are more interested in the B factor than the A factor.

Null and Alternate Hypotheses

The null and alternate hypotheses are the same as those for any two-way ANOVA. The null assumes no main effects and no interaction.

Partitioning the Variance

The total variability in a two-way repeated measures experiment can be partitioned in a similar manner to that used for the one-way design; this is illustrated below:

Two-way ANOVA–repeated measures

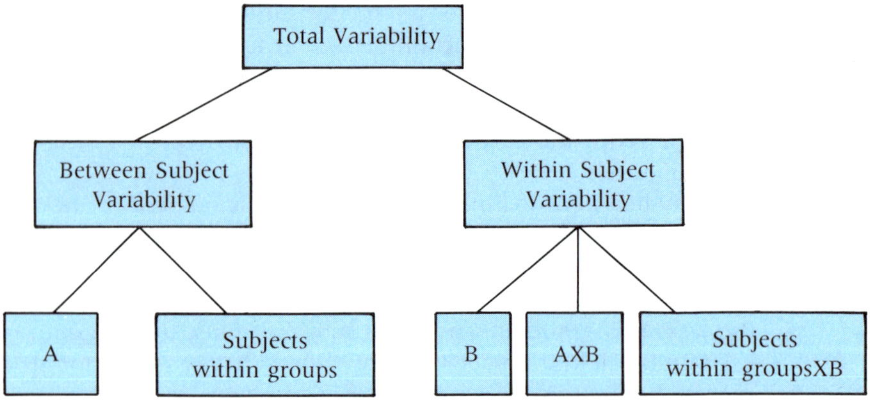

The total degrees of freedom can be partitioned as follows:

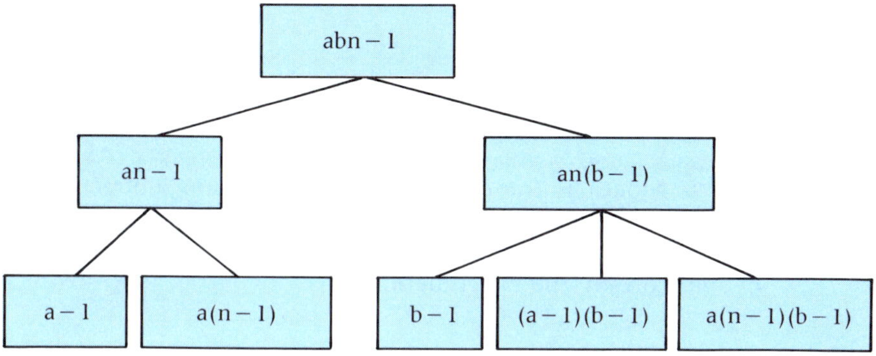

where a is the number of levels of factor A; b is the number of levels of factor B; and n is the number of subjects in each group.

Calculating the Sums of Squares

Below is an example of a 2x3 repeated measures design, with two levels of the between subject factor (A), and three levels of the within subjects factor (B), and a total of ten subjects in the experiment.

	Subject	B_1	B_2	B_3	Sum	
	1				S_1	
	2				S_2	
A_1	3				S_3	
	4				S_4	
	5				S_5	
	Sum	A_1B_1	A_1B_2	A_1B_3		A_1
	6				S_6	
A_2	7				S_7	
	8				S_8	
	9				S_9	
	10				S_{10}	
	Sum	A_2B_1	A_2B_2	A_2B_3		A_2
	Sum	B_1	B_2	B_3		

Step 1. Calculate the total sum of squares.

$$SS_{TOT} = \Sigma X_{tot}^2 - \frac{(\Sigma X_{tot})^2}{abn} \qquad \text{Sum of Squares Total}$$

We square all the raw scores, sum them and subtract from this value the square of the sum of all the scores divided by the total number of scores in the experiment. For a two-way repeated measures design, there is a total of abn scores in the experiment.

Step 2. Calculate the subject sum of squares (SS_S) and its component parts, the sum of squares for A(SS_A) and the sum of squares for subjects within groups, $SS_{S(gps)}$.

$$SS_S = \frac{(\Sigma S_1)^2 + (\Sigma S_2)^2 + \ldots + (\Sigma S_{an})^2}{b} - \frac{(\Sigma X_{tot})^2}{abn} \qquad \text{Sum of Squares Subjects}$$

The subject sum of squares is found by 1) summing across all levels of B for each subject in the experiment, 2) squaring each sum, 3) adding all these squared sums, 4) dividing the value in 3 by the number of levels of B, and 5) subtracting from the value in 4, the same value we used to obtain the total sum of squares in Step 1 (i.e., the squared sum of all the scores in the experiment divided by the total number).

This subject sum of squares can now be partitioned into its component parts. We first find the sum of squares for factor A.

$$SS_A = \frac{(\Sigma A_1)^2 + (\Sigma A_2)^2 + \ldots + (\Sigma A_a)^2}{bn} - \frac{(\Sigma X_{tot})^2}{abn} \qquad \text{Sum of Squares A}$$

This sum of squares is found by summing across all subjects and levels of factor B for each level of factor A, squaring each of these sums, summing the squares and dividing this value by the number of scores contributing to the sum (i.e., bn) and, finally, subtracting the same value as we used before.

The subjects within groups sum of squares can now be found by subtraction.

$$SS_{S(gps)} = SS_S - SS_A \qquad \text{Sum of Squares S(groups)}$$

Step 3. Calculate the within subject sum of squares and its component parts, the B, AXB, and subjects within groups XB sums of squares.

The within subject sum of squares is easily obtained by subtracting the subject sum of squares from the total sum of squares.

$$SS_{WS} = SS_{TOT} - SS_S$$

Now we compute the sum of squares for factor B.

$$SS_B = \frac{(\Sigma B_1)^2 + (\Sigma B_2)^2 + \ldots + (\Sigma B_b)^2}{an} - \frac{(\Sigma X_{tot})^2}{abn} \qquad \text{Sum of Squares B}$$

You can see that the procedure for calculating the sum of squares for factor B is much the same as that for factor A. We sum across all subjects and levels of A for each level of B, square each sum, add the squares and divide by the total number of subjects in each level of B (i.e., an).

Next, we compute the sum of squares for the AXB interaction.

$$SS_{AXB} = \frac{(\Sigma A_1B_1)^2 + (\Sigma A_1B_2)^2 + \ldots + (\Sigma A_aB_b)^2}{n} - \frac{(\Sigma X_{tot})^2}{abn} - SS_A - SS_B$$

The interaction sum of squares is found first by computing the sum for each combination of A and B. For a 3x2 design, we would have six sums. Then we add up the squares of all the sums, divide by the number of subjects in each group (i.e., n) and subtract from this value the same value we have used before. Then we subtract the SS_A and the SS_B.

Finally, we find the third component of the within subject sum of squares, the subjects within groups X B sum of squares ($SS_{S(gps)XB}$), by subtraction.

$$SS_{S(gps)XB} = SS_{WS} - SS_B - SS_{AXB}$$ Sum of Squares S(groups) XB

Calculating the Mean Squares and the F-Ratios

As with any two-way ANOVA, we will have three tests of significance, one for each main effect and one for the interaction.

The between subjects main effect of interest will be the A main effect; the mean square for A will be tested for significance with the mean square for subjects within groups (i.e., the error mean square for between subject comparisons).

The within subjects tests will be for the main effect of B and for the interaction of A and B. Both mean squares will be compared with the mean square for subjects within groups X B (i.e., the error mean square for the within subjects comparisons).

All mean squares are found by dividing the sums of squares by the appropriate degrees of freedom.

Between Subjects Mean Squares

$$MS_A = \frac{SS_A}{a - 1}$$ where a is the number of levels of factor A

$$MS_{S(gps)} = \frac{SS_{S(gps)}}{a(n - 1)}$$ where n is the number of subjects in each group

Within Subjects Mean Squares

$$MS_B = \frac{SS_B}{b - 1}$$ where b is the number of levels of factor B

$$MS_{AXB} = \frac{SS_{AXB}}{(a - 1)(b - 1)}$$

$$MS_{S(gps)XB} = \frac{SS_{S(gps)XB}}{a(n - 1)(b - 1)}$$

Now the three mean squares can be tested for significance.

F-Ratios for Two-Way ANOVA with Repeated Measures

$$F_A = \frac{MS_A}{MS_{S(gps)}}$$

$$F_B = \frac{MS_B}{MS_{S(gps)XB}}$$

$$F_{AXB} = \frac{MS_{AXB}}{MS_{S(gps)XB}}$$

Interpreting the F-Ratios

Each F-ratio is compared with the critical value from Table 3, by entering the appropriate degrees of freedom. Interpretation of the outcome of a two-way ANOVA is discussed in Chapter 11. A repeated measures design is interpreted in much the same way. Recall, however, that because subjects serve under all levels of B, tests of the within subjects treatment effects will be more sensitive.

Running a Two-Way ANOVA with Repeated Measures

Carla has been hired by the Police Department to assess the performance of its officers in several areas related to their jobs. The Department is primarily interested in three areas of competence: 1) general knowledge of police procedure, 2) general knowledge about the law and justice system in the province and 3) decision making ability in "real life" situations. The Department has a secondary interest in whether "time on the force" is an important factor.

Carla decides to use standardized tests available from the Police Department. All the multiple-choice tests are scored out of a maximum of 40 points. She randomly selects ten officers from each of four categories of "time on the job": 1) one year, 2) two to four years, 3) five to ten years, and 4) more than ten years on the Force.

All officers are tested in each of the three areas of competence. Carla, being a well-trained researcher, is concerned about the effects of previous testing on later test performance so she randomizes the order of the three tests in such a way that an equal number of officers within each group receive each order. In this way, she feels confident that the effects of repeated testing will be balanced among her groups. Carla's design is a 3x4 factorial design with repeated measures on factor B, type of test.

Here is her design.

Factor A Time on Force		Factor B: Type of Test		
		Procedure	Law	Decision Making
	Subject	B_1	B_2	B_3
	1			
	2			
	3			
	4			
A_1	5			
one year	6			
	7			
	8			
	9			
	10			
	11			
	12			
	13			
	14			
A_2	15			
2–4 years	16			
	17			
	18			
	19			
	20			
	21			
	22			
	23			
	24			
A_3	25			
5–10 years	26			
	27			
	28			
	29			
	30			
	31			
	32			
	33			
	34			
A_4	35			
>10 years	36			
	37			
	38			
	39			
	40			

And here are her data.

	B_1	B_2	B_3	sum		B_1	B_2	B_3	sum
	37	15	12	64		30	39	23	92
	28	20	11	59		22	33	27	82
	30	24	13	67		13	31	26	70
	22	25	12	59		15	25	15	55
A_1	33	21	14	68	A_3	17	30	18	65
	29	30	12	71		35	28	19	82
	27	35	13	75		33	22	18	73
	31	21	11	63		10	27	20	57
	36	22	11	69		18	32	24	74
	33	25	11	69		26	24	20	70
	25	28	13	66		26	22	23	71
	28	21	14	63		22	36	25	83
	30	30	17	77		21	33	28	82
	19	31	16	66		30	32	23	85
A_2	22	36	13	71		15	27	27	69
	29	22	13	64	A_4	18	25	27	70
	31	15	26	72		21	37	29	87
	13	20	13	46		27	20	26	73
	15	18	14	47		22	21	29	72
	20	30	15	65		16	17	27	60

Let's sum the scores for the subjects within each combination of A and B.

	B_1	B_2	B_3	Sum
A_1	306	238	120	664
A_2	232	251	154	637
A_3	219	291	210	720
A^4	218	270	264	752
Sum	975	1050	748	2773

Step 1. Calculate the total sum of squares.

$$SS_{TOT} = \Sigma X_{tot} - \frac{(\Sigma X_{tot})^2}{abn}$$

$$= 37^2 + 28^2 + \ldots + 27^2 - 2773^2/120$$

$$= 70349 - 64079.41$$

$$= 6269.59$$

Step 2. Calculate the SS_S and its component parts, SS_A and $SS_{S(gps)}$.

$$SS_S = \frac{(\Sigma S_1)^2 + (\Sigma S_2)^2 + \ldots + (\Sigma S_{40})^2}{b} - \frac{(\Sigma X_{tot})^2}{abn}$$

$$= \frac{64^2 + 59^2 + \ldots + 60^2}{3} - 64079.41$$

$$= 65349.00 - 64079.41 = 1269.59$$

$$SS_A = \frac{(\Sigma A_1)^2 + (\Sigma A_2)^2 + (\Sigma A_3)^2 + (\Sigma A_4)^2}{bn} - \frac{(\Sigma X_{tot})^2}{abn}$$

$$= \frac{664^2 + 637^2 + 720^2 + 752^2}{30} - 64079.41$$

$$= 64352.30 - 64079.41 = 272.89$$

$$SS_{S(gps)} = SS_S - SS_A = 996.70$$

Step 3. Calculate SS_{WS} and its component parts, the B, AXB, and subjects within groups XB sums of squares.

$$SS_{WS} = SS_{TOT} - SS_S = 5000.00$$

$$SS_B = \frac{(\Sigma B_1)^2 + (\Sigma B_2)^2 + (\Sigma B_3)^2}{an} - \frac{(\Sigma X_{tot})^2}{abn}$$

$$= \frac{975^2 + 1050^2 + 748^2}{40} - 64079.41$$

$$= 65315.73 - 64079.41 = 1236.32$$

$$SS_{AXB} = \frac{(\Sigma A_1B_1)^2 + (\Sigma A_1B_2)^2 + \ldots + (\Sigma A_4B_3)^2}{n} - \frac{(\Sigma X_{tot})^2}{abn} - SS_A - SS_B$$

$$= \frac{306^2 + 238^2 + \ldots + 264^2}{10} - 64079.41 - 272.89 - 1236.32$$

$$= 67208.30 - 64079.41 - 272.89 - 1236.32 = 1619.68$$

$$SS_{S(gps)XB} = SS_{WS} - SS_B - SS_{AXB}$$

$$= 2144.00$$

Step 4. Calculate the mean squares.

$$MS_A = \frac{SS_A}{a - 1} = 272.89/3 = 90.96$$

$$MS_{S(gps)} = \frac{SS_{S(gps)}}{a(n-1)} = 996.70/36 = 27.69$$

$$MS_B = \frac{SS_B}{b-1} = 1236.32/2 = 618.16$$

$$MS_{AXB} = \frac{SS_{AXB}}{(a-1)(b-1)} = 1619.68/6 = 269.95$$

$$MS_{S(gps)XB} = \frac{SS_{S(gps)XB}}{a(n-1)(b-1)} = 2144.00/72 = 29.78$$

Step 5. Calculate the F-ratios.

$$F_A = \frac{MS_A}{MS_{S(gps)}} = \frac{90.96}{27.69} = 3.29$$

$$F_B = \frac{MS_B}{MS_{S(gps)XB}} = \frac{618.16}{29.78} = 20.76$$

$$F_{AXB} = \frac{MS_{AXB}}{MS_{S(gps)XB}} = \frac{269.95}{29.78} = 9.07$$

Step 6. Enter the results in an ANOVA summary table.

ANOVA Summary Table

Source	SS	df	MS	F	p
Between Subjects	1269.59	39			
A	272.89	3	90.96	3.29	< .05
S (gps)	996.70	36	27.69		
Within Subjects	5000.00	80			
B	1236.32	2	618.16	20.76	< .01
AXB	1619.68	6	269.95	9.07	< .01
S (gps) XB	2144.00	72	29.78		
Total	6269.59	119			

Now, of course, we need to interpret these results. Since both main effects are significant, we know that type of test made a difference in performance and that length of time on the force made a difference in performance. However, the presence of a significant interaction tells us that the test difference pattern is not identical for the four groups. The best way I know of determining what went on in a design as complicated as this one is to present the data graphically.

Interpreting a Two-Way ANOVA with Repeated Measures

First we will look at the main effect of time on force.

Main effect of time on force

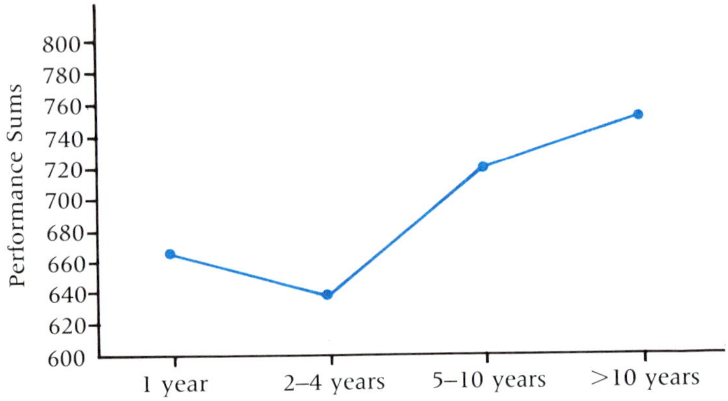

We can see that, overall, the more senior officers tended to do better; although the first year group did better than the 2–4 year group. In order to determine which groups are significantly different, this analysis would be followed by a multiple comparison technique as discussed in Chapter 13.

Let's look at the main effect of type of test.

Main effect of type of test

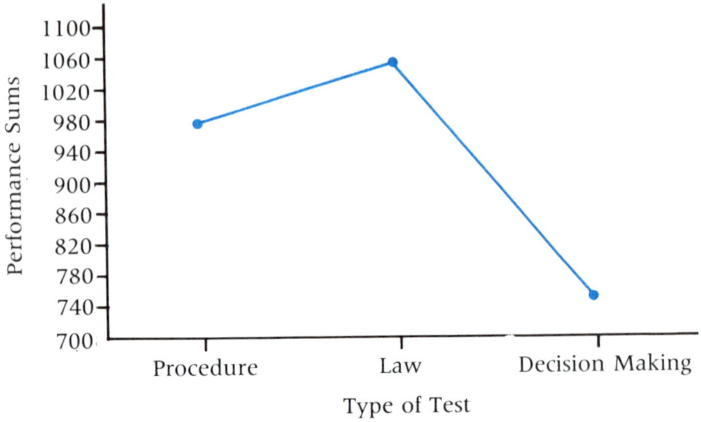

Overall, we see that the officers performed best on the law test and worst on the decision-making test.

Both these main effects must be interpreted somewhat cautiously because of the existence of a significant interaction between the variables. Let's look at the interaction.

Interaction of time on force and type of test

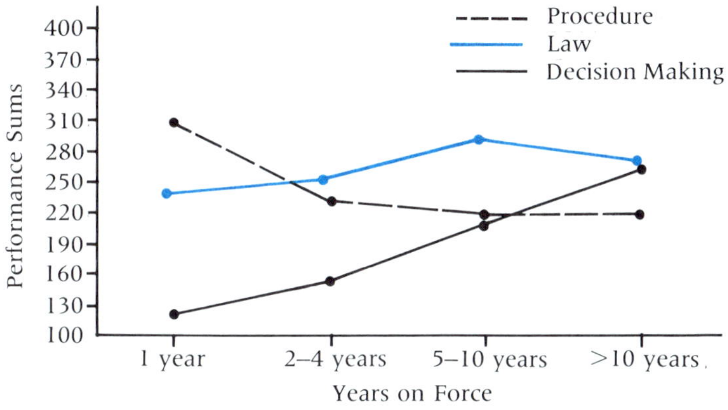

Examination of this chart indicates that as experience (i.e., years on the force) increases, performance on the decision making test increases also. The new officers did particularly well on the procedure test. We might speculate that the years of experience in the field improved the scores of the senior officers in "real life" decision making tests but that they tend to be a little complacent about matters involving procedure and law. We might also speculate that the new officers, on the other hand, lack real life experience but are keener about "book learning."

These speculative comments, however, should be supported by appropriate multiple comparison procedures.

Summary of Two-Way ANOVA with Repeated Measures

Hypotheses

H_o: No main effects and no interaction

H_a: H_o is false

Assumptions

1. Subjects are randomly selected with repeated measures on factor B.
2. Population distributions are normal.
3. Population variances are homogeneous.
4. Population covariances are equal.

Decision Rules

$df_a = a - 1 \quad df_b = b - 1 \quad df_{axb} = (a - 1)(b - 1)$

$df_{s(gps)} = a(n - 1) \quad df_{s(gps)xb} = a(n - 1)(b - 1)$

If $F_{obt} \geqslant F_{crit}$, reject the H_o

If $F_{obt} < F_{crit}$, do not reject H_o

Formulas

$$F = \frac{MS_A}{MS_{S(gps)}}$$

$$F_B = \frac{MS_B}{MS_{S(gps)XB}}$$

$$F_{AXB} = \frac{MS_{AXB}}{MS_{S(gps)XB}}$$

Assumptions Underlying ANOVA with Repeated Measures

One of the assumptions of ANOVA is that the populations from which the samples are drawn are normally distributed with equal variances. We discussed this assumption of homogeneity of variance in Chapter 11, where we learned that the F-test is quite robust (i.e., insensitive to violations of this assumption) when the subjects have been randomly selected and independently assigned to groups, and when sample sizes are similar.

With repeated measures designs, subjects are not independently assigned to groups and so the assumption of homogeneity takes on added importance. Another assumption of ANOVA is that the population correlations (often called covariances) are constant. If subjects are independently assigned to groups, it is reasonable to assume that they are constant. When the same subjects participate in all treatment levels, however, this is unlikely to be true. When there is reason to believe that both assumptions (homogeneity of variances and covariances) have been violated, a more conservative test is recommended (e.g., Geisser & Greenhouse, 1958).[1] Discussions of the problems of homogeneity of variances and covariances can be found in most upper-level statistics textbooks.

SUMMARY OF FORMULAS

One-Way ANOVA with Repeated Measures—Computational Formulas

Sums of Squares

Total

$$SS_{TOT} = \Sigma X_{tot}^2 - \frac{(\Sigma X_{tot})^2}{kn}$$

Between Subject

$$SS_S = \frac{(\Sigma S_1)^2 + (\Sigma S_2)^2 + \ldots + (\Sigma S_n)^2}{k} - \frac{(\Sigma X_{tot})^2}{kn}$$

Within Subject

$$SS_{WS} = SS_{TOT} - SS_S$$

Treatment

$$SS_T = \frac{(\Sigma T_1)^2 + (\Sigma T_2)^2 + \ldots + (\Sigma T_k)^2}{n} - \frac{(\Sigma X_{tot})^2}{kn}$$

Subject by Treatment

$$SS_{SXT} = SS_{WS} - SS_T$$

Mean Squares

Treatment

$$MS_T = \frac{SS_T}{k - 1}$$

Subject by Treatment

$$MS_{SXT} = \frac{SS_{SXT}}{(n - 1)(k - 1)}$$

F-Ratio

$$F = \frac{MS_T}{MS_{SXT}}$$

[1] Geisser, S. and Greenhouse, S.W. An extension of Box's results on the use of the F distribution in multivariate analysis. *Annals of Mathematical Statistics*, 1958, 29, 885–891.

Two-Way ANOVA with Repeated Measures — Computational Formulas

Sums of Squares

Total

$$SS_{TOT} = \Sigma X_{tot}^2 - \frac{(\Sigma X_{tot})^2}{abn}$$

Between Subjects

$$SS_S = \frac{(\Sigma S_1)^2 + (\Sigma S_2)^2 + \ldots + (\Sigma S_{an})^2}{b} - \frac{(\Sigma X_{tot})^2}{abn}$$

A

$$SS_A = \frac{(\Sigma A_1)^2 + (\Sigma A_2)^2 + \ldots + (\Sigma A_a)^2}{bn} - \frac{(\Sigma X_{tot})^2}{abn}$$

Subjects within Groups

$$SS_{S(gps)} = SS_S - SS_A$$

B

$$SS_B = \frac{(\Sigma B_1)^2 + (\Sigma B_2)^2 + \ldots + (\Sigma B_b)^2}{an} - \frac{(\Sigma X_{tot})^2}{abn}$$

AXB

$$SS_{AXB} = \frac{(\Sigma A_1B_1)^2 + (\Sigma A_1B_2)^2 + \ldots + (\Sigma A_aB_b)^2}{n} - \frac{(\Sigma X_{tot})^2}{abn} - SS_A - SS_B$$

Subjects within Groups XB

$$SS_{S(gps)XB} = SS_{WS} - SS_B - SS_{AXB}$$

Mean Squares

A

$$MS_A = \frac{SS_A}{a - 1}$$

Subjects within Groups

$$MS_{S(gps)} = \frac{SS_{S(gps)}}{a(n - 1)}$$

B

$$MS_B = \frac{SS_B}{b - 1}$$

AXB

$$MS_{AXB} = \frac{SS_{AXB}}{(a - 1)(b - 1)}$$

Subjects within Groups XB

$$MS_{S(gps)XB} = \frac{SS_{S(gps)XB}}{a(n - 1)(b - 1)}$$

F-Ratios

$$F_A = \frac{MS}{MS_{S(gps)}}$$

$$F_B = \frac{MS_B}{MS_{S(gps)XB}}$$

$$F_{AXB} = \frac{MS_{AXB}}{MS_{S(gps)XB}}$$

EXERCISES

1. Below is a layout for a one-way ANOVA with repeated measures.

Subjects	Treatments				
1	1	2	3	4	5
•					
•					
•					
20					

How many degrees of freedom for

a) Total SS
b) Subject SS
c) Within subject SS
d) Treatment SS
e) Subject by treatment SS

2. Run a one-way ANOVA at $\alpha = .05$ on the following data. All subjects served under each treatment condition.

	Subject		Treatments	
	1	2	3	Sum
1	22	34	21	77
2	35	36	20	91
3	31	40	17	88
4	17	41	23	81
5	22	33	27	82
6	20	36	29	85
7	35	28	31	94
8	21	29	22	72
9	33	34	15	82
10	10	30	28	68
Sum	246	341	233	820

3. Complete the ANOVA summary table for the following. There were 4 levels of A and 3 levels of B with 9 subjects under each level of A. Use $\alpha = .05$.

	Sums of Squares
Total	3500
Subject	250
A	50
S(gs)	200
ws	3250
B	700
AXB	550
S(gps)XB	2000

4. A psychology professor is interested in mid-term exam performance of his two sections of Introductory Psychology. He randomly selects 8 students from each section. He wants to compare their performance on the three midterm exams he uses during the course. Run a two-way ANOVA with repeated measures on test performance at $\alpha = .01$ on the following data.

Section A	Test 1	Test 2	Test 3
1	88	87	90
2	76	75	80
3	65	70	75
4	40	68	65
5	76	75	70
6	78	84	80
7	40	56	60
8	80	83	79
Section B			
1	75	75	80
2	72	78	80
3	56	63	70
4	69	74	70
5	63	76	75
6	71	70	72
7	73	69	69
8	64	69	78

5. Graph the two main effects on the interaction from Exercise 4. Use group means on your graphs.

CHAPTER 13

Multiple Comparison Procedures

In Chapters 11 and 12, we learned that the outcome of ANOVA tells us whether or not a significant difference exists. It does not tell us where that difference lies. ANOVA must be followed by additional statistical procedures to determine which means are different.

Controlling the Error Rate

One of the important concerns in doing inferential statistics is the possibility of making an error. A Type I error, you will recall, is made when we reject a null hypothesis that should not have been rejected. ANOVA keeps the Type I error rate at alpha over the entire experiment. If, on the other hand, we do multiple t-tests, rather than ANOVA, the Type I error rate is higher than alpha. Although the probability of a Type I error for each individual t-test is alpha, the probability of making at least one Type I error with many t-tests within one experiment is higher than alpha. The **experiment-wise** error rate is higher than the error rate for each t-test comparison. The experiment-wise error rate will depend on the number of t-test comparisons that are made within a single experiment. Since researchers must control the overall experiment-wise error rate, other procedures are recommended for comparing group means following ANOVA. Some common multiple comparison procedures are discussed in the following sections.

The Scheffé Method

The Scheffé method is appropriate for making any and all comparisons on a set of means. This method is considered superior to some of the other techniques when complex comparisons are of interest and/or when sample sizes are not equal. When samples are the same size and when simple comparisons are of interest, the Scheffé method is more conservative (less powerful) than some of the other techniques.

Constructing a Comparison

The Scheffé method allows comparisons between any and all means. In any comparison, two quantities are contrasted. These quantities may both be sample means or averages of means. For example, a researcher who used a control group and two experimental groups may wish to compare the control group mean with the average of the experimental group means. This could be expressed as

$$\overline{X}_c \text{ versus } \frac{\overline{X}_1 + \overline{X}_2}{2}$$

Or, he may wish to compare the two experimental group means with each other: $\overline{X}_1$ versus $\overline{X}_2$.

To construct a comparison, each quantity is multiplied by a coefficient. In the first example, $+1$ is the coefficient for the control mean and $-1/2$ is the coefficient for each experimental group mean. Since the null hypothesis is at the population level, for this example, it is

$$H_o\text{: } 1\mu_c - \frac{1}{2}\mu_1 - \frac{1}{2}\mu_2 = 0$$

For the second example, the coefficients are $+1$ and -1 and the null hypothesis is

$$H_o\text{: } 1\mu_1 - 1\,\mu_2 = 0$$

Alternate hypotheses are always non-directional.

In general, any comparison (C) may be expressed as

$$C = c_1\overline{X}_1 + c_2\overline{X}_2 + \ldots + c_k\overline{X}_k \qquad \text{A Comparison}$$

where c is the coefficient for each mean and k is the number of groups.

Note that the sum of the coefficients must be zero, i.e., ($c_1 + c_2 + \ldots + c_k = 0$).

Group means not included in the comparison are assigned coefficients of zero.

The Standard Error of a Comparison

To test the significance of a comparison, we need to determine the standard error(s_c). The general formula is

$$s_c = \sqrt{MS_{error}\left(\frac{c_1^2}{n_1} + \frac{c_2^2}{n_2} + \ldots + \frac{c_k^2}{n_k}\right)} \quad \text{Standard Error of a Comparison}$$

where MS_{error} is the appropriate error variance estimate from the ANOVA; c is the coefficient for the group mean; and n is the sample size.

Evaluating a Comparison for Significance

To evaluate a comparison for significance, we first must calculate F_s, the critical value of F used for a Scheffé comparison.

$$F_s = \sqrt{(k - 1)F_{crit}} \quad \text{Scheffé's Critical F}$$

where k is the number of groups.

F_{crit} is the tabled value of F with (k − 1) df in the numerator and the degrees of freedom associated with the MS_{error} from the Analysis of Variance, in the denominator.

Now we can go ahead and test our comparison for significance. The formula is

$$F' = C/s_c \quad \text{Scheffé F' Statistic}$$

As usual, if the obtained F value for any comparison is equal to or larger than the critical value, we reject the null hypothesis.

Running a Scheffé Test

Let's use an example from Chapter 11 to see how Scheffé's test works.

We ran an ANOVA to see if training students in study and test-taking skills made a difference in their performance. A control group received no training. There were 15 subjects in each group and their mean performance was

$\bar{X}_1$ (Control) = 59.07 $\quad$ $\bar{X}_2$ (Study) = 66.40 $\quad$ $\bar{X}_3$ (Test) = 71.73

The outcome of the ANOVA was

Source of Variance	Sum of Squares	df	Mean Squares	F	p
Between groups	1213.33	2	606.60	5.61	< .01
Within groups	4545.47	42	108.23		
Total	5758.80	44			

The ANOVA told us that there was a significant difference between the three kinds of students: those trained in study skills, those trained in test-taking skills and those receiving no training. Six comparisons can be examined: S vs T, S vs C, T vs C, ST vs C, TC vs S, SC vs T. Let's run a Scheffé test to determine whether the training made a difference. In other words we will compare the control group with the two trained groups. The null hypothesis will be: $H_0: 1/2\ \mu_2 + 1/2\ \mu_3 - 1\ \mu_1 = 0$.

Step 1. Construct the comparison

$$C = c_1\bar{X}_1 + c_2\bar{X}_2 + c_3\bar{X}_3$$

$$C = \left(\frac{1}{2}\right)(66.40) + \left(\frac{1}{2}\right)(71.73) + (-1)(59.07)$$

$$C = 69.07 - 59.07 = 10$$

Step 2. Determine the Standard Error

$$s_c = \sqrt{MS_{WG}\left(\frac{c_1^2}{n_1} + \frac{c_2^2}{n_2} + \frac{c_3^2}{n_3}\right)}$$

$$= \sqrt{108.23\left(\frac{\left(\frac{1}{2}\right)^2}{15} + \frac{\left(\frac{1}{2}\right)^2}{15} + \frac{(-1)^2}{15}\right)}$$

$$= 3.29$$

Step 3. Determine Scheffé's Critical F.

$$F_s = \sqrt{(k - 1)F_{crit}}$$

F_{crit} is the tabled value with k − 1 df in the numerator and the df associated with the MS_{wG} (i.e., the error mean square from ANOVA) in the denominator.

$$F_{2,\ 42} \text{ at } \alpha = .05 = 3.22$$

$$F_s = \sqrt{2(3.22)} = 2.54$$

Step 4. Test the comparison for significance.

$$F' = C/s_c = 10/3.29 = 3.04$$

Step 5. Make a decision. Since the obtained F value is larger than the critical value, we reject the null and claim that the training significantly improved performance.

The Scheffé method is particularly appropriate when complex combinations of sample means are being compared. Another common technique is called the Tukey method.

Summary of the Scheffé Test

Hypotheses

H_o: each comparison is zero.

H_a: each comparison is not zero.

Assumptions

The outcome of the ANOVA was significant.

Decision Rules

If $F' \geqslant F_s$, reject the H_o

If $F' < F_s$, do not reject H_o

Formula

$$F' = C/s_c$$

$$F_s = \sqrt{(k - 1)F_{crit}}$$

The Tukey Method

The Tukey method is more powerful than Scheffé's for comparing pairs of means. However, it is less powerful for comparing combinations of means. The Tukey method compares the difference between each pair of means with a value called the **honestly significant difference** (HSD). The value of HSD is found by

$$\text{HSD} = q(\alpha, df_{error}, k)\sqrt{MS_{error}/n}$$

where q is the value from Table 5; df_{error} is the degrees of freedom associated with the MS_{error} from ANOVA; k is the number of groups; and n is the number of subjects within a group.

The value of q is found by entering Table 5 with the α level, the df associated with the MS_{error} from the ANOVA, and the number of groups involved in the analysis. The differences between pairs of means can then be compared with the value of HSD. If any difference is greater than or equal to the value of HSD, the two means are significantly different or "honestly significantly different."

As with the Scheffé test the null hypotheses specify no difference between the population means and the alternate hypotheses are non-directional.

Running a Tukey Test

Let's use the same example we used earlier to run a Tukey test at $\alpha = .05$ on each pair of means. The three samples had 15 observations each and their means were $\overline{X}_1$ (Control) = 59.07, $\overline{X}_2$(Study) = 66.40, and $\overline{X}_3$ (Test) = 71.73.

The outcome of the ANOVA was as follows:

Source of Variance	Sum of Squares	df	Mean Squares	F	p
Between groups	1213.33	2	606.60	5.61	< .01
Within groups	4545.47	42	108.23		
Total	5758.80	44			

Step 1. Determine the difference between each pair of group means.

Study − Control = 66.40 − 59.07 = 7.33
Test − Control = 71.73 − 59.07 = 12.66
Test − Study = 71.73 − 66.40 = 5.33

Step 2. Compute the value of HSD.

$$\text{HSD} = q(\alpha, df_{wG}, k)\sqrt{MS_{wg}/n}$$
$$= q(.05, 42, 3)\ \sqrt{108.23/15}$$
$$= 3.44(2.69) = 9.25$$

Step 3. Compare differences with the value of HSD and make a decision.

Since the only mean difference larger than HSD is that between the Test group and the Control group, we conclude that test training had a significant effect on performance compared to no training.

Several other techniques for multiple comparisons have been developed but the Scheffé and Tukey methods are probably the most commonly used because of their generality and utility. Those of you wishing to learn more about multiple comparison techniques will find useful descriptions in Winer (1971)[2] and Kirk (1968).[3]

Summary of the Tukey Test

Hypotheses

H_o: no difference between population means.
H_a: H_o is false.

Assumptions

The outcome of the ANOVA was significant.

Decision Rules

Any mean difference $\geqslant$ HSD, reject the H_o

Formula

$$\text{HSD} = q(\alpha, df_{error}, k)\ \sqrt{MS_{error}/n}$$

[2] Winer, B.J. *Statistical Principles in Experimental Design.* (New York: McGraw-Hill, 1971)

[3] Kirk, R.E. *Experimental Design: Procedures for the Behavioral Sciences.* (Belmont, Cal.: Brooks/Cole, 1968)

SUMMARY OF TERMS AND FORMULAS

Scheffé's test is suitable for comparing any or all pairs of means as well as complex combinations of means. Samples need not be the same size. The Scheffé test requires computation of the F′ statistic for each comparison of interest. Each F′ statistic is then compared to Scheffé's critical F value for significance.

The **Tukey** method is used for comparing pairs of means and requires samples to be the same size. For simple comparisons, the Tukey method is considered to be a more powerful test. The Tukey test compares the mean difference between each pair of sample means with the value of the honestly significant difference (**HSD**). Any mean differences equal or larger than the HSD are considered to be significant.

Test	**Formulas**	
Scheffé	$F' = C/s_c$	F′ Statistic
	$F_s = \sqrt{(k-1)F_{crit}}$	Critical F value
	$C = c_1\overline{X}_1 + c_2\overline{X}_2 + \ldots + c_k\overline{X}_k$	Comparison
	$s_c = \sqrt{MS_{error}\left(\frac{c_1^2}{n_1} + \frac{c_2^2}{n_2} + \ldots + \frac{c_k^2}{n_k}\right)}$	Standard Error
Tukey	$HSD = q(\alpha, df_{error}, k)\sqrt{MS_{error}/n}$	

EXERCISES

1. The outcome of a one-way ANOVA is provided below. Run a Scheffé test at $\alpha = .05$ for the following null hypotheses.

a) $1\,\mu_1 - 1\,\mu_2 = 0$
b) $1\,\mu_3 - 1/2\,\mu_1 - 1/2\,\mu_2 = 0$
c) $1\mu_3 - 1\,\mu_2 = 0$

Source	df	MS	F	p
Between Groups	2	32.60	7.36	< .05
Within Groups	12	4.43		

Means: Group 1 = 4.60 $\quad n_1 = n_2 = n_3 = 5$
Group 2 = 3.00
Group 3 = 8.00

2. Run a Tukey test on the following data for all possible comparisons at $\alpha = .05$.

S#	Control	Group A	Group B
1	6	1	4
2	4	3	4
3	9	3	5
4	10	2	6

$MS_{WG} = 14.13$

3. For the following data, complete the ANOVA summary table and run Scheffé's test for all possible comparisons ($\alpha = .05$).

Source	SS	df	MS	F
Between	28.40	2		
Within	9.20	6		

$\overline{X}_1 = 6.30$
$\overline{X}_2 = 1.70$
$\overline{X}_3 = 2.40$
$n_1 = n_2 = n_3 = 3$

4. A kayak manufacturer measures the time to complete a race course by professional racers in 4 different types of kayaks for 6 trials each. Here are the data.

Eclipse	Mirage	Dancer	Mark IV
1.4	1.7	2.0	3.0
1.2	1.8	2.1	3.1
1.0	1.9	2.0	2.9
1.6	2.2	2.4	2.8
1.8	2.4	2.6	3.4
1.0	2.7	2.8	3.5

a) Run a one-way ANOVA on the data above at $\alpha = .05$.
b) Compare the Eclipse with the Mirage using Scheffé's test ($\alpha = .05$). Do they differ?
c) Compare the Eclipse with the average of the other three boats using Scheffé's test ($\alpha = .05$). What is your conclusion?
d) Compare all possible means with Tukey's test ($\alpha = .05$). What are your conclusions?

CHAPTER **14**

Inference with the Chi-Square Distribution

Non-parametric techniques: Used to compare population distributions rather than parameters.

When we make inferences about the mean of a population or the difference between means of two or more populations, for example, we are using parametric techniques. We are inferring the values of population parameters. Some research questions do not involve specific parameters of a distribution but, rather, the entire frequency distribution. An automobile dealer might be interested in the relative popularity of six of her car models, for example. If her data showed that one model was more popular with her customers than the others, she might use this information when ordering new cars from the manufacturer. In this example, the data are not mean scores, rather, they are the number of cars sold of each type that interests the dealer.

If we are interested in the nature of the entire population, we may find a **non-parametric** technique to be appropriate. The **chi-square test** (symbolized by χ^2) is a non-parametric test that allows us to make inferences about population frequencies from sample frequencies. Non-parametric tests are useful for analyzing data measured on nominal and ordinal scales, since they do not make as many assumptions as parametric tests.

Chi-square tests compare obtained sample frequencies with those expected according to the null hypothesis. Chi-square can be used with frequency data or with proportion data, since proportions can always be converted into frequencies. Chi-square can be used for discrete variables or for continuous variables which have been categorized into discrete intervals.

The chi-square test compares the frequencies obtained in a sample with those expected if the null hypothesis were true. The null hypothesis states the expected frequencies of sample data based on certain assumptions about the population from which the sample was drawn. Of course, we would not expect sample frequencies to be *exactly* equal to expected frequencies, even if the null were true. Sample frequencies will vary somewhat from their hypothesized values because of sampling fluctuation. The question is "How much variation between obtained sample frequencies and their expected values would likely occur if the null were true?" To answer this question, we need a sampling distribution.

The Chi-Square Distribution

Like the t- and F-distributions, the chi-square distribution is a family of distributions; the shape of each determined by degrees of freedom. The chi-square distribution is a relative frequency distribution based on discrepancies between obtained frequencies and their expected values. The value of chi-square will be smaller when the obtained and expected frequencies are similar and larger when they are not. If the hypothesized frequencies are not the true population frequencies, the discrepancies between obtained and expected values will be large and so will the value of chi-square. We need to discover what values of chi-square would occur with random sampling when the null is true. We can then compare our obtained chi-square value with this distribution of values and determine whether our outcome is a likely one or not according to the null hypothesis.

Constructing the Sampling Distribution of Chi-Square

The random sampling distribution of chi-square is constructed in the following manner:

Step 1. Randomly select a sample from a population of a discrete variable whose expected frequencies are known and record the frequency for each category.

Step 2. Subtract the expected frequency (E) for each category from the observed frequency(O). Square each difference and divide it by the associated expected, frequency i.e., $(O - E)^2/E$. Sum these values for all categories of the variable. This is the chi-square statistic.

Step 3. Return the sample to the population.

Step 4. Repeat the first three steps until all possible samples have been drawn from the population.

Step 5. Place the chi-square values in a relative frequency distribution.

This relative frequency distribution is the random sampling distribution of the chi-square statistic. It has certain properties.

Characteristics of the Chi-Square Distribution

Because the chi-square value is computed by squaring the differences between the observed and expected values, it can never be negative.

The null hypothesis states that the category frequencies in the population equal a set of values. A sample selected from that population should reflect that set of values. When the obtained frequencies are larger or smaller than those expected, the value of chi-square increases. For this reason, the region of rejection always appears in the upper tail of the distribution.

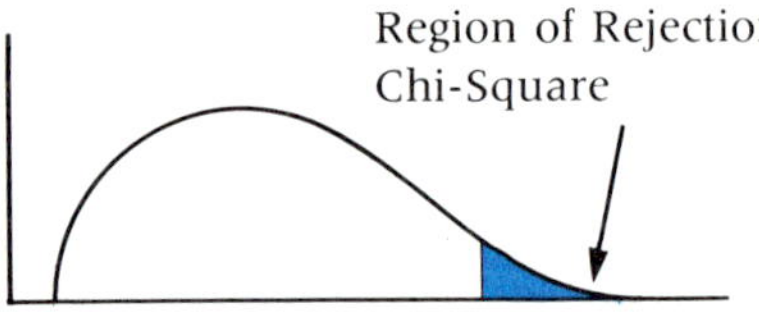

Although the critical region lies in one tail, the chi-square test is nondirectional. A very low chi-square value simply means that the obtained frequencies are closer to the expected values than chance would predict.

The chi-square distribution changes shape depending on degrees of freedom. In the other distributions we have looked at, degrees of freedom have been determined in relation to sample size. Degrees of freedom for the chi-square test are not related to sample size but rather are related to the number of discrepancies, (O − E)'s, that are independent and free to vary. In a chi-square test, we assume the total number of observations for total frequency is fixed and determine how many discrepancies are free to vary.

Using the Chi-Square Distribution

We will discuss the two chi-square tests: The **chi-square test for goodness of fit** and the **chi-square test for independence**. The first is used when we have one variable. The second test is used when we wish to discover whether two variables are related.

The Chi-Square Test for Goodness of Fit

This test is used when we have two or more categories or levels of one variable. We are interested in whether the frequencies we obtain for each category in our sample, match the expected frequencies specified by the null hypothesis.

Null and Alternate Hypotheses

The null and alternate hypotheses for a chi-square test of goodness of fit are determined before of data collection; that is *a priori*. How the null and alternate are stated depends upon the research question. For example, a tavern owner may be interested in customer preference for four types of beer. He may decide to have 100 customers taste each beer and indicate their preference. If the population, from which his customers came, had no preference for any of the four types, he would expect 25 people to choose each one. These would be the expected frequencies. The null hypothesis would specify "no difference in preference" and the expected value for each brand would be 25. The alternate hypothesis would state "preference differs" and that the true frequencies in the population were different than those hypothesized in the null.

H_o: No difference in preference for types of beer.

H_a: Preference for types of beer differs.

Calculating Chi-Square

Let's run a chi-square test for goodness of fit to answer the tavern owner's question. Suppose he finds, after 100 customers choose which of the four types of beer they prefer, the following frequencies.

	Types				
	Ale	Beer	Lager	Porter	Total
Observed Frequencies	10	20	55	15	100
Expected Frequencies	25	25	25	25	100

We find the value of chi-square as follows:

$$\chi^2 = \Sigma\frac{(O - E)^2}{E} \qquad \text{Chi-Square}$$

$$\chi^2 = \frac{(10 - 25)^2}{25} + \frac{(20 - 25)^2}{25} + \frac{(55 - 25)^2}{25} + \frac{(15 - 25)^2}{25}$$
$$= 50$$

Now he needs to compare the obtained χ^2 value with the critical value. The critical values for chi-square are found in Table 4 for various degrees of freedom. As usual, if the obtained value is equal to or larger than the critical value, the null hypothesis is rejected.

Recall that degrees of freedom for chi-square are determined by the number of comparisons between observed and expected frequencies that are free to vary. The total frequency is fixed. Looking at our example, we can see that three frequencies can vary but the fourth must be fixed to keep the sum at 100. Three discrepancies can vary, but the fourth cannot. One degree of freedom is used so the total can stay at 100. This problem, then, has three degrees of freedom.

The critical value of chi-square for 3df is 11.34 at $\alpha = .01$. Since the obtained value of 50 is greater than the critical value, the null is rejected. The tavern owner has statistical evidence of a significant difference in population preference for his four types of beer.

In this example the expected frequencies for each category were equal. This is not always the case. Remember Mendel? Imagine a botanist has determined that seedlings should show four particular characteristics in the ratio of 4:3:2:1, according to Mendel's Law. The first characteristic should appear in 4/10 of the seedlings, the second in 3/10, and so on. If the botanist collected a random sample of 200 seedlings, the expected frequencies for each characteristic would be 80, 60, 40, and 20. The obtained values could then be compared with these expected values.

Both of these examples have four categories and three degrees of freedom. A special situation exists when there are two categories and one degree of freedom. With only two categories of frequencies, either a chi-square test for goodness of fit of a Z-test for a proportion may be used, as long as the assumptions of Z have been met. The outcome will be the same. For example, suppose a researcher counts the number of gerbils who prefer a pink running wheel and the number who prefer a turquoise running wheel when given a choice between the two in their daily exercise period. Of the 50 gerbils in the experiment, 33 chose the pink wheel and 17 chose the turquoise wheel. The investigator wants to know if the color of the wheel affected their choice. If color makes no difference, we would expect equal numbers of gerbils to select each wheel. This experiment has two categories (pink and turquoise) and 1 degree of freedom; we may run a chi-square test or a Z-test for a proportion since our frequencies can be easily converted to proportion. Let's do both at $\alpha = .01$. The null and alternate hypotheses for a chi-square test are

H_o: Equal preference for pink and turquoise wheels
H_a: Preference differs

$$\chi^2 = \frac{(33 - 25)^2}{25} + \frac{(17 - 25)^2}{25} = 5.12$$

With 1df, the critical value of $\chi^2 = 6.64$ at $\alpha = .01$; we would not reject the null hypothesis. Gerbils, evidently, have no preference for the color of their running wheels.

With a Z-test for a proportion we hypothesize that the proportion of gerbils in the population who prefer pink running wheels is 0.50. In our sample, the proportion who chose the pink wheel was $33/50 = 0.66$

$$H_o: P = 0.50$$
$$H_a: P \neq 0.50$$

Then we conduct a Z-test for a proportion.

$$Z = \frac{p - P_p}{\sqrt{PQ/n}} = \frac{0.66 - 0.50}{\sqrt{(0.5)(0.5)/50}} = \frac{0.20}{0.07} = 2.29$$

The critical value of Z is ± 2.58 at $\alpha = 0.01$, so we cannot reject the null. There is no evidence that the proportion of gerbils preferring pink running wheels is different than 0.50.

Interpreting Chi-Square

The tavern owner interested in customer preference chose to test four types of beer. What if he had decided to test ten types? The degrees of freedom would be 9, not 3. The computed chi-square value would be larger because there were six more discrepancies in its calculation.

We found in the t- and F-distribution that as degrees of freedom increase, the critical value decreases. Perhaps, you are wondering why we don't just add more categories, increasing the value of our obtained chi-square, and making it easier to reject the null. A good point. If you examine the chi-square table, you will notice that as degrees of freedom increase, the critical value of chi-square *gets larger, not smaller.* Adding more categories does not make it easier to reject the null. The table takes the number of degrees of freedom into account when it determines the critical value. With more degrees of freedom in your experiment, you must obtain a larger chi-square value to reject the null.

The chi-square test for goodness of fit compares obtained frequencies with *a priori* expected values stated by the null hypothesis. The chi-square test for independence makes a different comparison.

Summary of the Chi-Square Test for Goodness of Fit

Hypotheses:

H_o: O's = E's

H_a: O's ≠ E's

Assumptions

1. Subjects are randomly selected.
2. Categories are mutually exclusive.

Decision Rules

df = # categories − 1

If $\chi^2_{obt} \geq \chi^2_{crit}$, reject H_o

If $\chi^2_{obt} < \chi^2_{crit}$, do not reject H_o

Formula

$$\chi^2 = \Sigma\frac{(O - E)^2}{E}$$

The Chi-Square Test for Independence

The chi-square test for independence is used to measure the association between two variables. This is often called the "two variable case." The question asked is "Are the two variables independent?" Suppose the tavern owner, discussed earlier, wished to know if preference for his four types of beer depended on whether the customer was female or male. His research question in this case would be "Are gender and beer preference independent?" Similarly, the gerbil researcher may wonder if preference for pink and turquoise running wheels depends on the previous experience of the gerbils in such wheels. His research question would be "Are preference for wheel color and previous experience independent?"

Null and Alternate Hypotheses

In the chi-square test for independence, the null hypothesis always states that the two variables are independent, and the alternate hypothesis states that they are dependent.

In the one-variable case (goodness of fit) expected frequencies were determined *a priori* by some theoretical assumption. In the two-variable case, the expected values are determined after data collection, or *post hoc*, from the values of the obtained or observed frequencies.

Determining Expected Frequencies

Let's call on our tavern owner again to illustrate how expected frequencies are determined in a chi-square test for independence. Suppose he randomly selected 150 men and 100 women and asked each to declare his or her preference for the four types of beer. Here are his data.

	Observed Frequencies				
	Ale	Beer	Lager	Porter	Total
Women	35	25	15	25	100
Men	20	30	70	30	150
Total	55	55	85	55	250

To find the expected frequencies for this problem, we must determine the frequencies we would expect in each category if gender made no difference to preference. If we examine the column totals we see that, overall, lager is the "preferred type" and the other three are ranked equally. If gender and preference are independent, then that pattern of results would be expected for both genders. In other words, if gender didn't matter, how many women would prefer each type and how many men would, considering the size of our two samples? The formula for determining these expected values is

$$E = \frac{(\text{Row Sum})(\text{Column Sum})}{\text{Total}}$$

Expected Values for the chi-Square test for independence

Using this formula, our expected frequencies for each group would be

	Expected Frequencies				
	Ale	Beer	Lager	Porter	Total
Women	22	22	34	22	100
Men	33	33	51	33	150
Total	55	55	85	55	250

Notice that the row and column totals are the same as those of the observed frequencies. This is always true. Now we are ready to compute our chi-square.

Calculating Chi-Square

$$\chi^2 = \Sigma\frac{(O - E)^2}{E}$$

$$\chi^2 = \frac{(35 - 22)^2}{22} + \frac{(25 - 22)^2}{22} + \frac{(15 - 34)^2}{34} + \frac{(25 - 22)^2}{22} +$$

$$\frac{(20 - 33)^2}{33} + \frac{(30 - 33)^2}{33} + \frac{(70 - 51)^2}{51} + \frac{(30 - 33)^2}{33} = 31.75$$

Now, we need to look up the critical value of chi-square and compare it to our obtained value. However, we need to know the degrees of freedom. Degrees of freedom are determined by the number of discrepancies that are free to vary. In a chi-square test for independence, we assume that the row and column sums are fixed. Degrees of freedom are equal to the number of cells in the table of expected values that are independent and, therefore, free to vary. One way to determine degrees of freedom is to ask how many expected values must be calculated before we can obtain the rest by subtraction? Let's look at our table of expected values, for a moment.

Expected Frequencies

	Ale	Beer	Lager	Porter	Total
Women	*				100
Men					150
Total	55	55	85	55	250

Let's calculate the expected frequency for the box marked with an asterisk.

$$E = \frac{(\text{Row Sum})(\text{Column Sum})}{\text{Total}} = \frac{(100)(55)}{250} = 22$$

We can find the expected value for men choosing ale by subtraction. It must be $55 - 22 = 33$. This value, then, is not free to vary. How many more values must we compute before we can obtain the rest by subtraction? Clearly, we need to calculate only two more values. Three expected values, altogether, must be calculated before the rest are fixed. Our problem, then, has 3df.

Summary of the Chi-Square Test for Independence

Hypotheses

H_o: The variables are independent.

H_a: The variables are dependent.

Assumptions

1. Subjects are randomly selected.
2. Observations have been classified simultaneously on two independent categories.

Decision Rules

df = (# rows − 1)(# columns − 1)

If $\chi^2_{obt} \geq \chi^2_{crit}$, reject H_o

If $\chi^2_{obt} < \chi^2_{crit}$, do not reject H_o

Formula

$$\chi^2 = \Sigma\frac{(O - E)^2}{E}$$

For any chi-square test of independence, degrees of freedom are calculated by:

$$df = (\# \text{ rows} - 1)(\# \text{ columns} - 1)$$

Returning to our problem:

$$df = (2 - 1)(4 - 1) = 3$$

The critical value of χ^2 is 7.82 at $\alpha = .05$. Our obtained value is larger than the critical value and so we reject the null hypothesis. We have evidence that gender and preference for four types of beer are dependent. In other words, women and men do not prefer the same types of beer.

Interpreting Chi-Square

A significant value of χ^2 means that the two variables are related in some systematic way. The overall pattern of each frequency distribution is not the same. Like the goodness of fit test, as categories are added to one or both variables, the χ^2 value will increase. The table takes this into account; the critical values increase with larger degrees of freedom.

Assumptions Underlying Inference with the Chi-Square Distribution

Chi-square assumes that the sample was randomly selected from the population, the observations are independent, and the variable discrete. Each particular observation can only be recorded in one cell. In other words, the categories are mutually exclusive.

In repeated experiments, chi-square requires that observed frequencies are normally distributed around their expected frequencies. This makes sense, because if we have an expected frequency of 15, occasionally we would observe a much higher or lower frequency, but most of the time we would see frequencies close to the expected value, given the null is true. This assumption may cause problems when expected frequencies are very small. With an expected frequency of 3, for example, the observed frequencies would be positively skewed around 3 since their distribution is limited at zero. Some statisticians recommend that all expected frequencies should be 5 or greater.

SUMMARY OF TERMS AND FORMULAS

The chi-squares tests are **non-parametric** analyses used to test hypotheses about frequencies of categorical or discrete variables. The test for **goodness of fit** is used to evaluate the discrepancy between observed and expected frequencies for categories of a single variable.

The test for **independence** evaluates the relationship between two variables with a null hypothesis that the variables are unrelated. Degrees of freedom for chi-square tests are determined by the number of discrepancies, used in the computation, that are independent and free to vary.

Chi-Square Formula

$$\chi^2 = \Sigma\frac{(O - E)^2}{E}$$

	Degrees of Freedom
Test for Goodness of Fit	number of categories − 1
Test for Independence	(# rows − 1)(# columns − 1)

EXERCISES

1. For each of the following, determine the degrees of freedom and the critical value of chi-square.

a) Test for goodness of fit, 6 categories, $\alpha = .05$
b) Test for goodness of fit, 2 categories, $\alpha = .01$
c) Test for independence, 4 categories for one variable and 5 categories for the second variable, $\alpha = .05$
d) Test for independence, 2 categories for each variable, $\alpha = .01$

2. A breakfast cereal manufacturer observed that in a random sample of 60 children, 27 preferred a cornflakes product, 19 preferred a shredded wheat product, and 14 preferred a high-fibre product. Use chi-square to test the hypothesis that there is no difference in children's preference for the three cereals. Use $\alpha = .05$.

3. A census determined that 60% of Canadians are Protestant, 30% are Catholic, and 10% have no religious affiliation. A small religious college of 500 students has 300 Protestants, 190 Catholics, and 10 students with no religious affiliation. Test the hypothesis at $\alpha = .05$, that the college students are a random sample of the general Canadian population.

4. When offered a choice between popsicles and ice cream, 35 children chose popsicles and 15 chose ice cream. At $\alpha = .01$, test the hypothesis that children's preference does not differ between the two treats.

5. A sporting goods manufacturer wants to determine if there is a relationship between gender and the riskiness of the sport each sex engages in. Over a six-month period, he records whether the equipment purchased is for a high-risk sport (kayaking, downhill skiing, skydiving) or a low-risk sport (cross-country skiing, skating, windsurfing) and whether the purchaser was male or female. He finds the following:

	High-Risk	Low-Risk
Men	22	37
Women	18	28

Test at $\alpha = .05$ whether riskiness and gender are independent.

6. A psychologist administers a test to 120 randomly selected churchgoers to assess the strength of their religious values. A second test is then administered to the same group to assess attitude toward legalization of marijuana. Use a chi-square test to see if the two variables are related at $\alpha = .01$. The data are as follows:

		Piety			
		High	Medium	Low	Total
Attitude	In Favor	5	10	20	35
	Neutral	15	20	10	45
	Against	20	10	10	40
	Total	40	40	40	

7. A sociologist wonders if "blondes have more fun." She selects a random sample of 50 blondes, 40 brunettes, and 35 redheads. She records the number of dates each girl has over a six-month period and classifies them into three categories. Test the hypothesis that hair color and "popularity" are independent, at = .05. The data are as follows:

Number of Dates	Blondes	Brunettes	Redheads	Total
>50	39	15	13	67
25–50	8	20	10	38
<25	3	5	12	20
Total	50	40	35	

8. A scientist has been experimenting with black gerbils. He claims that as a result of certain injections, when two black gerbils are mated, the offspring will be black, white and gray in the proportion 5:4:3. Many gerbils were mated after being injected with the chemical. Of 170 newborn gerbils, 61 were black, 69 were white, and 40 were gray. Test the scientist's claim at $\alpha = .01$.

9. Two kayaking buddies, Dan and Peter, enjoy racing each other down river. Over several years, they have been very evenly matched. Each has won about half the races. Last season, Peter read a book called "How to Win in Kayak Racing." Since then, Peter has lost 18 races against Dan, out of 20. Use chi-square, at $\alpha = .05$, to test the hypothesis that the book had no effect on Peter's performance.

10. A faculty member and his wife are getting ready to attend a garden party given by his university. He wants to dress casually, but his wife thinks he should wear a suit. He explains to her that while most of the administrative staff attending the party will be in suits, most of the faculty will not. He lost the argument, but when they arrived at the party, he made a careful tally of who (administrators vs faculty) were wearing what (suit or no-suit). At $\alpha = .01$, test the hypothesis that position and style of dress are independent. The data are as follows:

	Position at University		
	Administrator	Faculty	Total
Suit	45	8	53
No-Suit	10	32	42
Total	55	40	

CHAPTER 15

Additional Non-Parametric Techniques

Many research studies involve rank order, rating scale, or frequency count data. These kinds of data are usually not interval or ratio in scale, and so inferences about means, for example, are inappropriate because the underlying assumptions of parametric analyses cannot be met. When we have nominal or ordinal data, we must use procedures specifically developed for them. Chapter 13 presented the chi-square test, a non-parametric technique appropriate for frequency data. This chapter presents some additional non-parametric analyses.

The Mann-Whitney U Test

The **Mann-Whitney U test** is used to make inferences about the difference between two populations. The test is sensitive to the entire distributions from which the samples were drawn as well as their central tendencies.

This test, used for ordinal data, is the non-parametric alternative to the t-test for the difference between independent groups.

Null and Alternate Hypotheses

The null hypothesis states that the populations from which the samples were drawn are identical. The alternate hypothesis states they are different.

H_o: The populations are identical.

H_a: The populations are not identical.

If the population distributions are similar in shape, the test compares their central tendencies. If the central tendencies are similar the test compares the entire distributions in general. As you can see, these hypotheses are slightly different from those used in inference with the t-distribution.

The Mann-Whitney U test requires that subjects are randomly selected and independently assigned to groups and that the scores are ranked in order.

The U Statistic

The Mann-Whitney test computes the **U** statistic which follows the U distribution. The obtained U value is evaluated in terms of the sampling distribution of the U statistic. The U statistic which is tested for significance is the smaller of the following two values:

$$U_1 = n_1 n_2 + \frac{n_1(n_1 + 1)}{2} - \Sigma R_1$$

Mann-Whitney U Test

$$U_2 = n_1 n_2 + \frac{n_2(n_2 + 1)}{2} - \Sigma R_2$$

where n_1 is the sample size of group 1; n_2 is the sample size of group 2; ΣR_1 is the sum of the ranks for group 1; and ΣR_2 is the sum of the ranks for group 2.

The critical values of U are found in Table 6 for one- and two-tailed tests of significance. Enter the sample size for the first group along the top of the table, and the sample size of the second group along the side of the table. Unlike our previous tests, *the obtained U value must be smaller than the critical value to be significant.*

Running the Mann-Whitney U test

Step 1. Assign ranks to the scores in the groups.

This is done by combining and arranging the scores from both groups in order from the smallest to the largest and assigning a rank where 1 is the smallest score. For example, if a subject in Group 1 had the smallest score then that subject's rank would be 1. If the next smallest score was obtained by a subject in Group 2, then that subject would be assigned a rank of 2.

Step 2. Calculate the U statistic.

Step 3. Compare the obtained value with the critical value. If the obtained U value is smaller than the critical value, reject the null hypothesis, otherwise, do not reject the null hypothesis.

Ready for an example? Ten psychology professors and ten biology professors were given a questionnaire designed to measure their attitudes toward various controversial issues about the influence of heredity and environment on human behavior. Here are the data arranged in order for each group:

DATA: Scores on Questionnaire

Biology Professors	Psychology Professors
2	6
3	8
11	16
13	17
15	23
25	24
27	26
33	37
39	38
45	49

First we need to assign ranks to all the scores. The smallest score, 2, appears in the biology professors group and that score is assigned a rank of 1. Here are the rank order data:

DATA: Ranks

Biology Professors	Psychology Professors
1	3
2	4
5	8
6	9
7	10
12	11
14	13
15	16
18	17
19	20
Total = 99	Total = 111

Summary of the Mann-Whitney U Test

Hypotheses

H_o: Populations are identical.
H_a: Populations are not identical.

Assumptions

1. Subjects are randomly selected and independently assigned to groups.
2. Measurement scale is at least ordinal.

Decision Rules

If $U_{obt} \leq U_{crit}$, reject the H_o

If $U_{obt} > U_{crit}$, do not reject the H_o

Formula

$$U_1 = n_1 n_2 + \frac{n_1(n_1 + 1)}{2} - \Sigma R_1$$

$$U_2 = n_1 n_2 + \frac{n_2(n_2 + 1)}{2} - \Sigma R_2$$

U_{obt} is the smaller of U_1 and U_2

Now we can use the formulas to compute each U value.

$$U_1 = n_1 n_2 + \frac{n_1(n_1 + 1)}{2} - \Sigma R_1$$

$$= 10(10) + \frac{10(11)}{2} - 99$$

$$= 100 + 55 - 99 = 56$$

$$U_2 = n_1 n_2 + \frac{n_2(n_2 + 1)}{2} - \Sigma R_2$$

$$= 10(10) + \frac{10(11)}{2} - 111$$

$$= 100 + 55 - 111 = 44$$

The second U value is smaller, so we compare it to the critical value. At $\alpha = .05$, the critical value is 23 for a two-tailed test. Since the smaller obtained U value is larger than the critical value, the null hypothesis is not rejected. We have no evidence that biology and psychology professors differ in terms of their attitudes toward the effect of heredity and environment on behavior. Remember that for this test the obtained U must be smaller than the critical value before we can reject the null.

Running the Mann-Whitney U for Large Sample Sizes

The sampling distribution of the U statistic approaches the normal distribution when both samples have 20 or more observations. In such cases, we can compute a Z value and use the normal curve tables in our test of significance.

After ranks have been assigned to the scores we can run the Mann-Whitney U in the following way:

Step 1. Determine the mean and standard deviation of the U statistic.

The sampling distribution of the U statistic has a mean of

$$\mu_u = n_1 n_2/2$$

and a standard deviation of

$$\sigma_u = \sqrt{\frac{n_1 n_2(n_1 + n_2 + 1)}{12}}$$

Step 2. Compute the Z statistic.

Subtract the mean of the sampling distribution of our statistic (μ_u) from the obtained sample outcome (U) and divide by the standard error of the sampling distribution (σ_u). In notation:

$$Z = \frac{U - \mu_u}{\sigma_u} \qquad \text{Z formula for the U statistic}$$

Step 3. Compare the obtained Z value with the critical value. If the obtained value is equal to or larger than the critical value, reject the null hypothesis, otherwise, do not reject the null hypothesis.

Ready for an example? Twenty randomly selected women and twenty randomly selected men are given a questionnaire designed to assess their feelings about the status of women in the work place. The data are considered ordinal. Arranged from lowest to highest, here are the scores.

	Scores		Ranks	
	Women	Men	Women	Men
	3	5	1	3.5
	4	7	2	6
	5	9	3.5	8.5
	6	10	5	10
	8	11	7	11
	9	18	8.5	17
	12	19	12	18
	14	20	13.5	19.5
	14	21	13.5	21
	16	23	15	23
	17	24	16	24
	20	29	19.5	27
	22	31	22	28
	27	34	25	30
	28	37	26	33
	32	39	29	35
	35	41	31	36
	36	46	32	38
	38	48	34	39
	43	52	37	40
Totals			352.5	467.5

You will notice that some of the scores are equal. For example, two scores of 5 occurred, one in the women's group, the other in the men's. A common way to deal with tied scores is to assign the average of the ranks whose positions they take. Look at the two 5's that occurred. If these two scores were different, they would take positions 3 and 4. Because they are the same, they are assigned the average of those two rank positions, that is 3.5. If three 5's had occurred, each would be given the rank of 4, the average of positions 3, 4 and 5.

We can now go ahead and run our test. We first need to compute U_1 and U_2.

$$U_1 = n_1 n_2 + \frac{n_1(n_1 + 1)}{2} - \Sigma R_1$$

$$= 20(20) + \frac{20(21)}{2} - 352.5$$

$$= 257.5$$

$$U_2 = n_1 n_2 + \frac{n_2(n_2 + 1)}{2} - \Sigma R_2$$

$$= 20(20) + \frac{20(21)}{2} - 467.5$$

$$= 142.5$$

Since U_2 is the smaller value, we use 142.5 as the U value in our Z-test.

$$Z = \frac{U - \mu_u}{\sigma_u}$$

$$= \frac{U - n_1 n_2/2}{\sqrt{\dfrac{n_1 n_2(n_1 + n_2 + 1)}{12}}}$$

$$= \frac{142.5 - (20)(20)/2}{\sqrt{(20)(20)(41)/12}}$$

$$= \frac{-57.5}{36.97}$$

$$= -1.56$$

The critical value for a Z-test is ± 1.96 for a two-tailed test at the .05 level of significance. Since our obtained Z value was less than the critical value, we do not reject the null hypothesis. In other words, we have no evidence that men and women feel differently about the status of women in the workplace.

The Mann-Whitney test is the non-parametric alternative to the t-test for independent groups. It requires that the two samples be independent. If the two groups are dependent, for instance, the subjects are matched on certain characteristics or the groups contain the same subjects, then a different non-parametric test is required.

The Wilcoxon Signed-Ranks Test

The Wilcoxon test is used for ordinal data obtained on the same or matched subjects.

Null and Alternate Hypotheses

The null and alternate hypotheses for the Wilcoxon test are identical to those in the Mann-Whitney test.

H_o: The populations are identical.

H_a: The populations are not identical.

The T Statistic

The Wilcoxon test computes the T statistic. (Note: This is not the same as the t-statistic). The T statistic follows the sampling distribution of T.

The following procedure is used to conduct the Wilcoxon test.

Step 1. Calculate the difference between the two scores for each subject.

Step 2. Rank the absolute values (i.e., ignore the + or − sign of each value) of the difference scores from lowest to highest. Place the appropriate sign (+ or −) next to each rank.

Step 3. Sum the ranks with the less frequent sign. In other words, if there are fewer positive ranks than negative ranks, sum the values of the positive ranks. This is the T statistic.

Step 4. Compare the obtained value with the critical value.

Table 7 provides the critical values of T for one- and two-tailed tests at various levels of significance. As with the Mann-Whitney test, the obtained T value *must be less than the critical value* for rejection of the null hypothesis.

Ready for an example?

Running the Wilcoxon Signed-Ranks Test

A French professor rated the pronunciation accuracy of 10 randomly selected students based on a pre-test. Scores could range from 0 to 15, with 15 representing excellent pronunciation. He provided these students with instruction in French pronunciation and then re-rated their accuracy on a post-test. Let's run a two-tailed test at $\alpha = .05$.

Student	Pre-test	Post-test	Difference	Rank
1	10	10	0	dropped
2	4	7	−3	−3.5
3	11	8	3	3.5
4	9	10	−1	−1
5	2	6	−4	−5
6	6	1	5	6
7	8	2	6	7
8	12	5	7	8
9	3	12	−9	−9
10	4	2	2	2

Like the Mann-Whitney test, tied difference scores are assigned shared ranks. Notice that when the two scores are equal and the difference is 0, the pair of scores is dropped from the analysis and n, the number of paired scores, is reduced accordingly.

There are fewer negatives than positives, so we sum all the ranks with negative signs. Our T value is 18.5.

To determine the critical value, we enter the table with the number of pairs of scores used in the final analysis. In our example, we dropped one pair of scores and so $n = 9$. We find that the critical value of T for a two-tailed test at $\alpha = .05$ is 6. Since our obtained T value is greater than the critical value, we do not reject the null. Remember that, for this test, the obtained value must be smaller than the critical value to reject the null. There is no evidence that the professor's instruction affected the students' pronunciation accuracy.

Running the Wilcoxon Signed-Ranks Test for Large Sample Sizes

Like the Mann-Whitney test, when the samples are reasonably large (>20), the sampling distribution of T approaches the normal distribution and the Z statistic is used in the test for significance.

Step 1. Determine the mean and standard deviation of the T statistic.

The mean of the sampling distribution of T is

$$\mu_T = \frac{n(n + 1)}{4}$$

and the standard deviation is

$$\sigma_T = \sqrt{\frac{n(n + 1)(2n + 1)}{24}}$$

Step 2. Compute the Z statistic.

The formula for converting T to Z is the following.

$$Z = \frac{T - \mu_T}{\sigma_T}$$ Z formula for the T statistic

Step 3. Compare the obtained Z value with the critical value. If the obtained Z value is equal to or larger than the critical value, reject the null hypothesis, otherwise, do not reject the null.

Let's look at an example. In a dog show, 20 Golden Retrievers were rated, by impartial judges, before and after receiving obedience training.

Ratings were based on a scale from 0 to 5 with 5 representing excellent performance. Let's test the null hypothesis that training had no effect on the dog's behavior. Here are the total points received by each dog:

	Rating Before	After	Difference	Rank
Spot	34	40	−6	−9.5
Fido	23	30	−7	−12
Breeze	56	55	1	1.5
Sydney	46	50	−4	−4.5
Amber	33	40	−7	−12
Dudley	38	39	−1	−1.5
Murphy	13	25	−12	−16
Karma	26	30	−4	−4.5
Shuba	22	25	−3	−3
Beast	37	43	−6	−9.5
Boga	19	36	−17	−20
Debit	27	34	−7	−12
Queenie	45	50	−5	−7
Bear	33	47	−14	−18
Murgatroyd	29	38	−9	−15
Fifi	30	25	5	7
Gemini	20	15	5	7
Phoenix	25	38	−13	−17
Luckenbach	15	31	−16	−19
Alfonse	50	42	8	14

In this example there are fewer positively signed ranks than negative ranks, so we sum the positive ranks. Our obtained T value is 29.5, the sum of the four positive ranks.

To compute the Z value:

$$Z = \frac{T - \mu_T}{\sigma_T}$$

$$= \frac{T - (n)(n + 1)/4}{\sqrt{\frac{n(n + 1)(2n + 1)}{24}}}$$

$$= \frac{29.5 - 20(21)/4}{\sqrt{20(21)(41)/24}}$$

$$= \frac{-75.5}{26.79}$$

$$= -2.82$$

Summary of the Wilcoxon test

Hypotheses

H_o: Populations are identical
H_a: Populations are not identical

Assumptions

1. Subjects are randomly selected.
2. Same or matched subjects.
3. Measurement scale is at least ordinal.

Decision Rules

If $T_{obt} \leq T_{crit}$, reject the H_o
If $T_{obt} > T_{crit}$, do not reject the H_o

Formula

n = number of pairs with non-zero differences.
T is the sum of the absolute ranks with the less frequently appearing sign.

The critical Z at $\alpha = .01$ is ± 2.58; our obtained Z value is larger numerically than the critical value so we reject the null hypothesis and accept the alternate. Training had a significant effect on the ratings.

The Mann-Whitney and Wilcoxon tests are alternative analyses to t-tests. When the assumptions underlying inference with the t-distribution have not been met, for example, the data are ordinal in scale, then these tests are appropriate for testing hypotheses about the difference between two population distributions.

In Chapter 11, you learned how to use the F-distribution to test hypotheses about several population means. When the assumptions underlying ANOVA have not been met, an alternative analysis is appropriate. The nonparametric analog to the one-way ANOVA is the **Kruskal-Wallis test.**

The Kruskal-Wallis Test

The Kruskal-Wallis test is used for ordinal data when subjects have been randomly selected and independently assigned to groups. The calculations for this test are similar to those used in the Mann-Whitney test.

Null and Alternate Hypotheses

The null hypothesis of the Kruskal-Wallis test states that all populations have identical distributions. The alternate is that the populations are not identical.

H_o: Populations are identical.

H_a: Populations are not identical.

The H Statistic

The statistic computed in the Kruskal-Wallis test is the **H** statistic. The sampling distribution of the H statistic follows the chi-square distribution with k − 1 degrees of freedom, where k is the number of samples or groups in the experiment.

The computational formula for the H statistic is

$$H = \frac{12}{n_{tot}(n_{tot}+1)}\left[\frac{(\Sigma R_1)^2}{n_1} + \frac{(\Sigma R_2)^2}{n_2} + \ldots + \frac{(\Sigma R_k)^2}{n_k}\right] - 3(n_{tot}+1)$$

where n_{tot} is the total number of observations; ΣR is the sum of the ranks of the scores in the group; and k is the number of groups.

Running the Kruskal-Wallis Test

The following lists the steps for running the Kruskal-Wallis test.

Step 1. Assign ranks to the scores in the groups.

All the scores are rank ordered from the smallest to the largest regardless of group membership. Ties are treated in the usual way.

Step 2. Calculate the H statistic.

Step 3. Compare the obtained H value with the critical value of χ^2. If the obtained value is equal to or larger than the critical value, reject the null hypothesis, otherwise, do not reject the null.

Let's look at an example. The President of a small junior college has obtained teaching evaluation ratings of eight professors in each of three faculties, Arts, Science, and Education. We compute the Kruskal-Wallis H to test the hypothesis that the ratings for Arts, Science, and Education professors don't differ. Here are the data.

Arts professors	Science professors	Education professors
13	23	17
12	30	10
16	29	28
27	14	20
19	11	22
18	9	21
15	24	8
31	32	26

These evaluation scores must be combined and then ranked from smallest to largest. The smallest score was obtained by the seventh Education professor, so this score is assigned the rank of 1. Here are the rank data.

	Arts professors	Science professors	Education professors
	6	16	10
	5	22	3
	9	21	20
	19	7	13
	12	4	15
	11	2	14
	8	17	1
	23	24	18
Total	93	113	94

Using our formula for the H statistic:

$$H = \frac{12}{n_{tot}(n_{tot} + 1)}\left[\frac{(\Sigma R_1)^2}{n_1} + \frac{(\Sigma R_2)^2}{n_2} + \ldots + \frac{(\Sigma R_3)^2}{n_3}\right] - 3(n_{tot} + 1)$$

$$= \frac{12}{24(24 + 1)}\left(\frac{93^2}{8} + \frac{113^2}{8} + \frac{94^2}{8}\right) - 3(24 + 1)$$

$$= 0.02(3781.75) - 75 = 0.63$$

The obtained H value is compared with the critical value of χ^2 with $k - 1$ degrees of freedom. In our example, $k = 3$, and the critical value of χ^2 for 2 degrees of freedom, at $\alpha = .05$, is 5.99. The obtained value is smaller than the critical value, so we do not reject the null hypothesis. We have no evidence that student evaluations of professors in Arts, Science, and Education are different.

The tests we have been studying are the non-parametric alternatives to some of the inferential techniques discussed in previous chapters. These are used for testing hypotheses about the equality of populations when the assumptions of parametric analysis have not been met.

Many research questions ask about the relationship between variables. In the next chapter, you will learn about **Pearson's correlation test**, a procedure for determining the strength of association between two variables. An adaptation of this test, the **rank-order correlation test**, is used for ordinal variables which have been rank-ordered.

Summary of the Kruskal-Wallis Test

Hypotheses

$\mathbf{H_o}$: Populations are identical

$\mathbf{H_a}$: Populations are not identical

Assumptions

1. Subjects are randomly selected and independently assigned to groups.
2. Measurement scale is at least ordinal.

Decision Rules

$df = k - 1$

If $H_{obt} \geq \chi^2_{crit}$, reject the H_o

If $H_{obt} < \chi^2_{crit}$, do not reject the H_o

Formula

$$H = \frac{12}{n_{tot}(n_{tot} + 1)} \left[\frac{(\Sigma R_1)^2}{n_1} + \frac{(\Sigma R_2)^2}{n_2} + \ldots + \frac{(\Sigma R_k)^2}{n_k} \right] - 3(n_{tot} + 1)$$

The Rank-Order Correlation Test

When ordinal variables have been ranked and you wish to determine the relationship between the ranks, the rank-order correlation test, often called **Spearman's Rho**, is appropriate.

Null and Alternate Hypotheses

The null hypothesizes no correlation between the ranks. The alternate hypothesis may be non-directional (i.e., the correlation is not zero) or directional (i.e., the correlation is greater or less than zero).

$$H_o\text{: Rho} = 0$$

$$H_a\text{: Rho} \neq 0\text{, or Rho} > 0\text{, or Rho} < 0$$

The rho Statistic

To be consistent, we will differentiate between the correlation in the population (Rho) and the sample statistic (rho).

rho is calculated on rank data with the following formula.

$$\text{rho} = 1 - \frac{6\Sigma d^2}{n(n^2 - 1)}$$

Rank-Order Correlation Formula

where n is the number of paired ranks; and d is the difference between the paired ranks.

Running the Rank-Order Correlation Test

The following describes the steps for running the rank-order correlation test.

Step 1. Determine the difference between the ranks for each subject.

Step 2. Square each difference and sum them.

Step 3. Calculate the rho statistic.

Step 4. Compare the obtained rho value with the critical value.

If the obtained value is equal to or larger than the critical value, reject the null hypothesis; otherwise, do not reject the null. Table 8 provides the critical values of rho for various sample sizes and levels of significance. Remember that the sample size is the number of *pairs* of ranks.

Ready for an example?

The Free Women's Coalition wishes to investigate the relationship between judges' rankings of entrants in the bathing suit competition and the talent competition in the Miss Canada Pageant. Here are the data:

	Rank-Orders			
Miss	Bathing Suit	Talent	d	d^2
Edmonton	1	1	0	0
Vancouver	2	2	0	0
Regina	3	3	0	0
Winnipeg	4	4	0	0
Montreal	5	7	−2	4
Quebec	6	6	0	0
P.E.I.	7	5	2	4
Ottawa	8	9	−1	1
Toronto	9	8	1	1
Calgary	10	10	0	0
				10

Using our formula:

$$\text{rho} = 1 - \frac{6\Sigma d^2}{n(n^2 - 1)}$$

$$= 1 - \frac{6\,(10)}{10(100 - 1)}$$

$$= 1 - \frac{60}{990}$$

$$= 1 - 0.06 = 0.94$$

The critical value of rho is 0.65, at $\alpha = 0.05$ for a two-tailed test, when there are ten subjects. The obtained value is larger than the critical value and the null hypothesis is rejected. There is a significant correlation between judges' rankings of the women in the bathing suit and talent competition.

The rank-order correlation test is very useful for ranked data. If the original data are not in rank order form, they must be converted to use the rho formula.

Summary of the Rank-Order Correlation Test

Hypotheses

H_o: Rho = 0
H_a: Rho ≠ 0 or Rho < 0 or Rho > 0

Assumptions

1. Subjects are randomly selected.
2. Observations are rank-ordered.

Decision Rules

n = number of pairs of ranks
If $rho_{obt} \geqslant rho_{crit}$, reject H_o
If $rho_{obt} < rho_{crit}$, do not reject H_o

Formula

$$rho = 1 - \frac{6\Sigma d^2}{n(n^2 - 1)}$$

SUMMARY OF TERMS AND FORMULAS

When the underlying assumption of parametric analyses have not been met, a **non-parametric** approach may be appropriate.

The **Mann-Whitney U test** is analogous to the t-test for the difference between **independent** means.

For samples which are correlated or **dependent**, the **Wilcoxon test** is appropriate and is the non-parametric equivalent to the t-test for dependent means.

When the assumptions of **ANOVA** have not been met, the **Kruskal-Wallis test** may be used.

Spearman's rank-order correlation test is used to determine the relationship between two sets of rank ordered data.

Test	Computational Formula
Mann-Whitney	$U_1 = n_1n_2 + \frac{n_1(n_1 + 1)}{2} - \Sigma R_1$
	$U_2 = n_1n_2 + \frac{n_2(n_2 + 1)}{2} - \Sigma R_2$
Z Formula	$\frac{U - n_1n_2/2}{\sqrt{n_1n_2(n_1 + n_2 + 1)/12}}$
Wilcoxon Z formula	$\frac{T - n(n + 1)/4}{\sqrt{n(n + 1)(2n + 1)/24}}$
Kruskal-Wallis	$H = \frac{12}{n_{tot}(n_{tot} + 1)}\left[\frac{(\Sigma R_1)^2}{n_1} + \frac{(\Sigma R_2)^2}{n_2} + \ldots + \frac{(\Sigma R_k)^2}{n_k}\right] - 3(n_{tot} + 1)$
Rank-Order Correlation	$rho = 1 - \frac{6\Sigma d^2}{n(n^2 - 1)}$

EXERCISES

1. A random sample of students from a Liberal Arts College and another random sample from a Church Work Training College are given a questionnaire designed to measure their attitude toward capital punishment. The scores are given below. A high score reflects pro capital punishment.

Liberal Arts Students	Church Work Students
10	6
11	5
12	3
14	7
18	6
8	2
9	4
7	10

Run a Mann-Whitney U test on these data with $\alpha = .01$. What are your conclusions?

2. A social psychologist used an Aggression Scale to rate the aggressiveness of children before and after they viewed a violent cartoon show. Run a Wilcoxon test on the data below at $\alpha = .05$. What are your conclusions?

Aggressiveness Rating

Before	After
2.0	2.5
3.5	2.5
1.5	2.0
2.5	2.5
3.0	1.5
4.0	4.5
1.0	2.5
3.0	2.5
1.5	3.0

3. Kayakers have been asked to rate the quality of 3 different boats. Run a Kruskal-Wallis test to see if the ratings differ ($\alpha = .05$). Do they?

Ratings

Eclipse	Mirage	Dancer
6.0	5.3	3.5
6.5	5.4	3.3
5.2	4.7	3.6
4.8	3.1	2.9
6.1	3.9	4.0

4. The hypothetical rank orders of the 10 contestants in the Miss Canada pageant for the last 2 years are provided below. Run the rank-order correlation test to see if the rankings are related ($\alpha = .05$). Are the two sets of ratings similar?

Miss	Last Year	This Year
B.C.	1	3
Alta.	2	1
Man.	3	5
Sask.	4	4
Ont.	5	2
Que.	6	8
P.E.I.	7	6
N.S.	8	7
N.B.	9	9
N.F.	10	10

5. Plot the data from question 4 on a scattergraph.

CHAPTER **16**

Pearson's Correlational Technique

Correlational techniques are used to determine if two variables are related. In Chapter 2 you learned how to construct a scattergraph of a bivariate frequency distribution. A scattergraph shows the relationship between two variables. By examining a scattergraph, you can see whether or not two variables appear to be related. This relationship can be quantified by a *coefficient of correlation*, a value which indicates the strength of the relationship between two variables.

Correlation is a valuable statistical technique used in a great deal of research, particularly in medical research. For example, the general health of pregnant women is correlated with the birth weight of their babies. Stress is correlated with heart problems in men. Did you know that men's hair loss is correlated with their educational level? As number of years of education increase, so does the amount of hair lost in male adults. This latter example illustrates a very important rule in correlation, *correlation does not infer causation*! The correlation between two variables does not tell us that one variable caused the other, or vice versa. A correlation, no matter how strong, does not provide any information about cause. A strong correlation only tells us that two variables tend to vary together. Why they do this is a question that only a well controlled *experiment* can answer.

When we say that two variables are positively correlated, we mean that the values of these variables tend to increase and decrease together. For example, studying is positively correlated with grades. In other words, if you increase your study time your grade point average will tend to increase as well.

When two variables are negatively correlated, the values of the first variable tend to increase as the values of the second variable decrease. Age and visual acuity are negatively correlated. As people get older, their vision tends to get poorer.

The correlation between high school grades and post-secondary performance may be of interest to you. After many years of collecting data, most colleges and universities have found that their students with higher high school grades tend to do better than those with poorer grades. In other words, there is a positive correlation between high school grades and college grades. Colleges and universities use this correlation for screening applicants to their programs. We will look at how correlational information is used to predict future events from current events in the next chapter.

Correlational techniques can be used not only to *describe* the relationship between two variables, but also to *infer* the correlation between two variables in a population based on a sample drawn from that population. We will start by discussing correlation as a descriptive technique and follow with correlation as an inferential technique.

Correlation as a Descriptive Technique

When two variables are correlated, the values of the first variable tend to increase or decrease in a regular fashion with the values of the second variable. When high values of one variable are associated with high values of another, and low values of the first are associated with low values of the second, they are *positively* correlated. On the other hand, when high values of one variable are associated with low values of another, they are *negatively* correlated.

The *correlation coefficient* tells us how strongly the two variables are related and in what fashion. All correlation coefficients range from +1 to −1. A correlation coefficient of 1 is a perfect correlation and means that values of one variable are exactly related to values of another variable. A correlation coefficient of 0 indicates that the values of the first variable are unrelated to the values of the second variable.

Several ways of computing correlation coefficients have been developed. The choice of one over another depends upon the variables of interest. When both variables are continuous, Pearson's product-moment coefficient of correlation is used.

Pearson's Product-Moment Coefficient of Correlation (ρ)

Pearson developed this coefficient by fitting a straight line to a bivariate frequency distribution. The line is called the **straight line of best fit**. The next chapter describes how Pearson fit this line to the data.

Basically, the coefficient of correlation (ρ, pronounced "rho") is a number describing how close the points in a bivariate frequency distribution fit a straight line. If all the points fall on the line, the correlation is 1: a perfect correlation. If all the points do not fall on the line, the correlation coefficient will be less than 1.

Calculating the Coefficient of Correlation (ρ)

The formula for calculating Pearson's coefficient of correlation varies depending on the data. We will look at versions of the formula for three types of data: raw scores, deviation scores, and Z-scores.

Data in Raw Score Form

For raw data, the formula for Pearson's ρ is

$$\rho = \frac{\Sigma XY - (\Sigma X)(\Sigma Y)/N}{\sqrt{[\Sigma X^2 - (\Sigma X)^2/N][\Sigma Y^2 - (\Sigma Y)^2/N]}} \qquad \text{For Raw Data}$$

In the numerator ΣXY is the sum of the *cross-products* of the values for each variable; $(\Sigma X)(\Sigma Y)$ is the product of the sum of X and the sum of Y; and N is the number of pairs of scores.

In the denominator, the square root is taken of the product of the sum of squares in X and the sum of squares in Y. Ready for an example? Suppose I gathered data about the television habits of ten children in my son's kindergarten class. I asked their teachers to rate the aggressiveness of each child, on a scale from 1 to 10, during each recess period for one month. Variable X is the mean rating per week for each child. I also asked the parents of these children to determine the amount of violent T.V. their children watch each week. The mean number of hours watched by each child per week is my Y variable. I am interested in determining whether those children who are rated as most aggressive are also the children who watch a lot of violent television programs. Let's determine whether these two variables, aggressiveness and T.V. habits, are related.

Child	Rating X	T.V. Y	X^2	Y^2	XY
Ben	8.5	5.8	72.3	33.6	49.30
Gabo	4.3	3.6	18.5	13.0	15.48
Linda	5.6	4.0	31.4	16.0	22.40
Luke	2.2	1.5	4.8	2.3	3.30
Hiro	6.5	6.3	42.3	39.7	40.95
Matthew	8.2	7.5	67.2	56.3	61.50
Mia	9.7	8.0	94.1	64.0	77.60
Ashley	1.0	2.2	1.0	4.8	2.20
Serge	3.5	3.1	12.3	9.6	10.85
Padraic	5.0	3.0	25.0	9.0	15.00
Totals	54.5	45.0	368.8	248.2	298.58

$$\rho = \frac{\Sigma XY - (\Sigma X)(\Sigma Y)/N}{\sqrt{[\Sigma X^2 - (\Sigma X)^2/N][\Sigma Y^2 - (\Sigma Y)^2/N}}$$

$$= \frac{298.58 - (54.5)(45)/10}{\sqrt{[368.8 - (54.5)^2/10][248.2 - (45)^2/10]}}$$

$$= \frac{53.33}{(8.47)(6.76)} = 0.93$$

The correlation between teachers' ratings of the children's aggressiveness and number of hours watching violent T.V., recorded by the parents, is 0.93. This suggests that the children who are rated most aggressive watch a lot of violent television programs, according to their parents.

Data in Deviation Score Form

When the data are in deviation score form, the formula for Pearson's coefficient is

$$\rho = \frac{\Sigma xy}{\sqrt{(\Sigma x^2)(\Sigma y^2)}} \qquad \text{For Deviation Data}$$

Where Σxy is the sum of the cross-products of the deviation scores for each variable; (Σx^2) is the sum of the squared deviations in X; and (Σy^2) is the sum of the squared deviations in Y.

Let's use the previous example to verify our formula. Here are the data in deviation score form.

	x	y	x^2	y^2	xy
Ben	3.0	1.3	9.0	1.7	3.9
Gabor	−1.2	−0.9	1.4	0.8	1.1
Linda	0.1	−0.5	0.0	0.3	−0.1
Luke	−3.3	−3.0	10.9	9.0	9.9
Hiro	1.0	1.8	1.0	3.2	1.8
Matthew	2.7	3.0	7.3	9.0	8.1
Mia	4.2	3.5	17.6	12.3	14.7
Ashley	−4.5	−2.3	20.3	5.3	10.3
Serge	−2.0	−1.4	4.0	2.0	2.8
Padraic	−0.5	−1.5	0.3	2.3	0.8
Total			71.77	45.74	53.33

Using the deviation score formula:

$$\rho = \frac{\Sigma xy}{\sqrt{(\Sigma x^2)(\Sigma y^2)}}$$

$$\rho = \frac{53.33}{\sqrt{(71.77)(45.74)}} = \frac{53.33}{57.3} = 0.93$$

As you can see, the result is the same as that for the raw score data.

Data in Z-Score Form

When the data are in Z-score form, the formula for Pearson's correlation is

$$\rho = \frac{\Sigma(Z_x Z_y)}{N} \qquad \text{For Z-Score Data}$$

The cross-products of the Z-scores are summed and divided by the number of scores.

To show how this formula works, we will use the same data we used before. Now the scores have been converted to Z-scores.

	Z_x	Z_y	Z_xZ_y
Ben	1.12	0.61	0.68
Gabor	−0.45	−0.42	0.19
Linda	0.04	−0.23	−0.01
Luke	−1.23	−1.40	1.73
Hiro	0.37	0.84	0.31
Matthew	1.01	1.40	1.41
Mia	1.57	−1.64	2.56
Ashley	−1.68	−1.07	1.80
Serge	−0.75	−0.65	0.49
Padraic	−0.19	−0.70	0.13
Total			9.30

$$\rho = \frac{\Sigma(Z_xZ_y)}{N} = \frac{9.30}{10} = 0.93$$

Once again, the result is the same because this formula is algebraically equivalent to the others.

Factors Influencing the Correlation Coefficient

I am sure every statistics student and most introductory psychology students have been told over and over that *correlation does not infer causation.* You know that the coefficient of correlation measures the degree to which two variables are related. It does not tell you anything about the reason for the relationship. If variable A is correlated with variable B, then we know that values of A tend to be associated with values of B. We do not know if A causes B, if B causes A, or if some other variable is responsible for the relationship. Earlier the correlation between hair loss and educational level of males was mentioned. It appeared that as men become more educated, they tend to lose more hair. Now if we were naive, we might think that education *causes* hair loss; perhaps studying causes stress which causes hair loss. Or perhaps hair loss *causes* men to seek more education; bald men do not date much, so they have more time to fill, so they take more courses. Clearly this is ridiculous. What we are actually seeing is a relationship between two variables, hair loss and education, which is caused by a third variable, age. Older men tend to have more years of education. Similarly, because they are older, they have less hair! Consider the following statements:

"People who brush with 'Brand X' toothpaste have fewer cavities."

"People who use 'Brand X' detergent have whiter clothes."

These statements imply that the use of a product will cause certain wonderful things to happen. This is not necessarily the case.

Linearity of Regression

The strength of the correlation depends on how closely the points hug the straight line of best fit. In a particular set of data, however, a straight line may or may not describe the relationship between the variables. A straight line is appropriate for variables that are linearly related. But what if X and Y are not linearly related?

Below are scattergraphs of two bivariate frequency distributions.

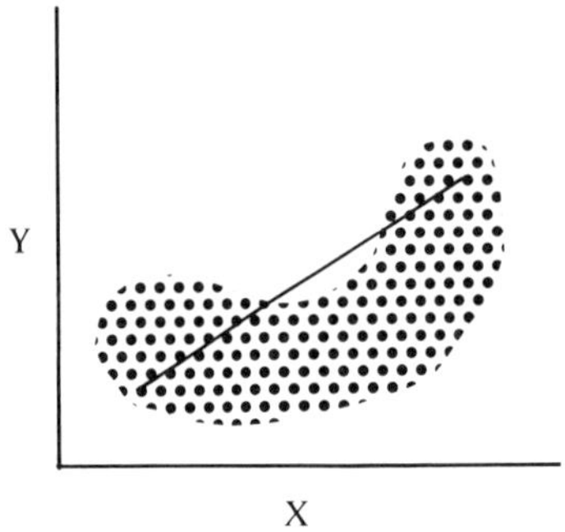

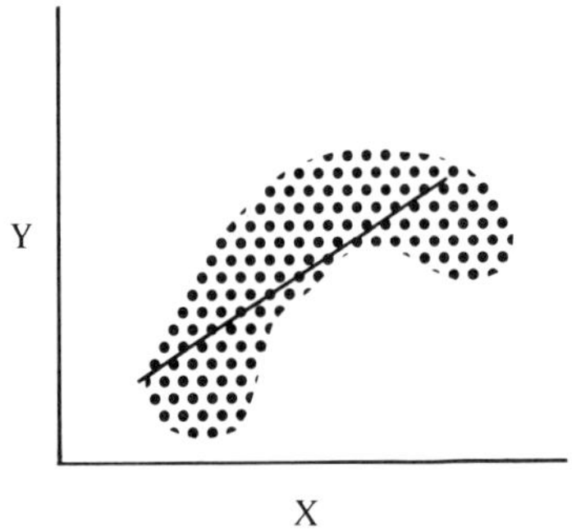

If we calculated the correlation coefficient, ρ, for either of these distributions, it would be quite low; the points do not fit a straight line very well. In both cases, a curvilinear function best describes the relationship between the variables. Whenever the relationship between two variables is not linear, ρ, will *underestimate* the strength of the relationship.

Homoscedasticity

When the amount of scatter is the same throughout a bivariate distribution, it is said to have the property of homoscedasticity or equal variability.

Examine the two scattergraphs below.

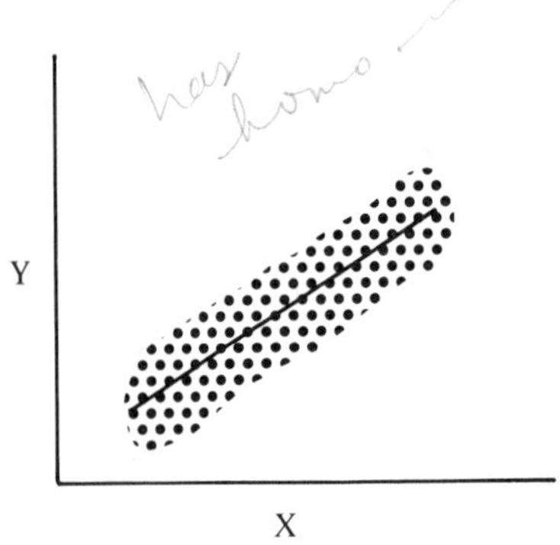

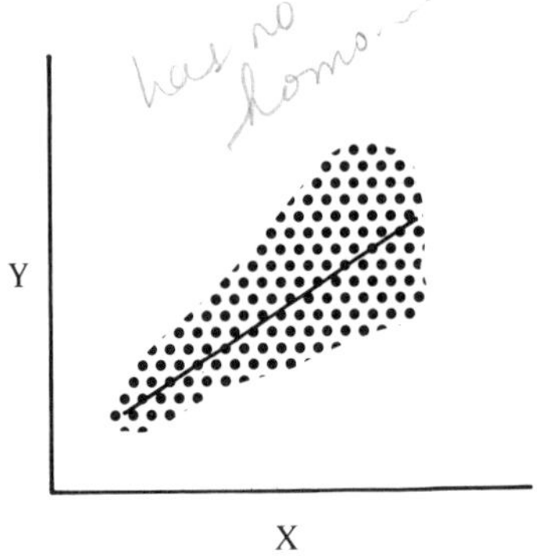

The scattergraph on the left shows homoscedasticity. No matter which value of X is chosen, the corresponding Y values are equally scattered around the straight line of best fit. In the scattergraph on the right, however, this is not the case. If X is low, Y does not vary much around the line; for higher values of X, the corresponding Y values vary a great deal. The second distribution does not have homoscedasticity.

Pearson's ρ reflects the "average" degree to which the scores hug the line of best fit. For the scattergraph on the right, ρ will underestimate the strength of the relationship for low values of X and overestimate it for high values of X.

Discontinuous Distributions

Whenever the range of one or both variables is restricted in some way, ρ will be affected. For example, if the middle of a distribution is excluded, the ρ will be higher than if the middle had been included. This might occur if we were to correlate the high and low scores without including the middle scores. Suppose the Dean of Arts wished to know if the grade point average of his 1st year Arts students correlated with their high school grades. He has data for students who failed and for those who received honors in their first year. He has no data for the middle students. If he calculates Pearson's ρ for this *discontinuous* distribution, the ρ value will be somewhat higher than if he had all the data available. Similarly, if the extreme scores were omitted from the calculation, ρ will tend to be lower.

Interpreting the Coefficient of Correlation

Coefficient of Determination: Indicates the strength of relationship between two variables.

Unless ρ is 1, the coefficient of correlation does not directly indicate the association between Y and X. For example, a ρ value of 0.50 does not mean that there is a 50% association between the two variables. However, the square of ρ, does indicate the degree of the association.

The coefficient of determination, ρ^2, reflects the degree of association between Y and X. For a correlation coefficient of 0.50, the coefficient of determination (ρ^2) is 0.25. This indicates the strength of the association. Specifically, 0.25 or 1/4 of the total Y variance is due to the correlation of Y with X. The coefficient of determination will be discussed in detail in the next chapter. For now, remember that ρ^2 reflects the amount of association between the two variables.

Correlation as an Inferential Technique

When we draw samples from a population, determine the correlation between two variables, and make inferences about the value of the correlation in the population, we are using correlation as an inferential technique. Like any inferential technique, we do this when we can't obtain all the observations in a population.

We will examine one test involving correlation — testing the hypothesis that the population correlation is zero. As you will see, the formulas and notation change when we use correlation in inference.

Null and Alternate Hypotheses

The null hypothesis states that, in the bivariate population from which the sample was selected, the correlation is zero.

$$H_o: \rho = 0$$

The alternate hypothesis may be directional or non-directional.

$$H_a: \rho \neq 0$$
$$H_a: \rho > 0$$
$$H_a: \rho < 0$$

You will recall that whenever we sample from a population, the sample statistics vary from sample to sample. This is called sampling fluctuation. Even if the correlation in a population was truly zero, we would not expect a sample correlation of exactly zero. We would expect some variability. Our question here, as with any inferential technique, is "How far from zero would our sample correlation be expected to vary if the true population correlation is indeed zero?"

To answer this question, we need a distribution of correlations based on samples drawn from a population with a correlation of zero; the **random sampling distribution of the correlation coefficient**. The mean of this sampling distribution will be zero. The coefficient of correlation computed on a sample is called "r." The estimate of the standard deviation or standard error of this sampling distribution is denoted as s_r. When the null hypothesis is true, the sampling distribution of the correlation coefficient is close to normal for reasonably large samples.

Testing the Significance of the Correlation

To test the significance of a correlation, we need to compute r from the sample data. The formula is

$$r = \frac{\Sigma XY - (\Sigma X)(\Sigma Y)/n}{\sqrt{[\Sigma X^2 - (\Sigma X)^2/n][\Sigma Y^2 - (\Sigma Y)^2/n]}} \quad \text{r for a Sample}$$

We can covert our sample statistic, r, to a standard score. As you recall, this is done by subtracting the hypothesized parameter from the sample statistic and dividing the difference by the standard error of the sampling distribution of the statistic.

$$t = \frac{r - \rho}{s_r} \quad \text{t Formula for the r Statistic}$$

The estimate of the standard deviation, the standard error, is found by the following formula:

$$s_r = \sqrt{\frac{1 - r^2}{n - 2}} \quad \text{Standard Error for the r Statistic}$$

When we do this for r, we obtain standard scores which follow the t-distribution with $n - 2$ degrees of freedom, where n refers to the number of pairs of scores.

After computing the t value, we test our coefficient for significance by looking up the critical value of t for $n - 2$ degrees of freedom.

Let's do an example. A researcher randomly selects ten students and administers a test of creativity and a test of mathematical competence. He discovers that the correlation between the scores on the two tests is 0.80. He wants to test the hypothesis that creativity is significantly correlated to mathematical competence.

$H_o: \rho = 0$
$H_a: \rho \neq 0$

$$s_r = \sqrt{\frac{1 - r^2}{n - 2}} = \sqrt{\frac{1 - 0.80^2}{8}} = 0.21$$

$$t = \frac{r - \rho}{s_r} = 3.77$$

The critical value of t is ± 2.31 with 8df at $\alpha = .05$. Since the obtained t falls in the region of rejection, the null hypothesis is rejected. The researcher has evidence that creativity and mathematical competence are significantly correlated.

Assumptions Underlying Inference about Correlations

The assumptions underlying tests of significance of correlation coefficients are similar to those of several parametric analyses we have discussed previously. First, the subjects are assumed to be randomly selected from the population. Second, normal population distributions of both X and Y are assumed. When the sample is reasonably large (30 or more), this assumption will not be badly violated because the sampling distribution of r tends to be normal, regardless of population shape.

Summary of the Pearson Correlation Test

Hypotheses

H_o: $\rho = 0$
H_a: $\rho \neq 0, \rho < 0, \rho > 0$

Assumptions

1. Subjects are randomly selected.
2. Both populations are normally distributed.

Decision Rules

If $t_{obt} \geq t_{crit}$, reject H_o
If $t_{obt} < t_{crit}$, do not reject H_o

Formula

$$t = \frac{r - \rho}{\sqrt{\frac{1 - r^2}{n - 2}}}$$

SUMMARY OF TERMS AND FORMULAS

Pearson's product-moment correlation coefficient is used to determine the extent of relationship between two variables. The numerical size of the coefficient indicates the strength and the sign indicates the direction of the relationship. The square of the coefficient, the **coefficient of determination**, indicates how much of the total variance in the Y dimension is accounted for by the association of Y with the X variable.

Pearson's technique is suitable for **describing** the relationship between two variables and for making **inferences** from sample data about the correlation in the population.

Pearson's ρ	**Formulas**
For Raw Data	$\dfrac{\Sigma XY - (\Sigma X)(\Sigma Y)/N}{\sqrt{[\Sigma X^2 - (\Sigma X)^2/N][\Sigma Y^2 - (\Sigma Y)^2/N]}}$
For Deviation Data	$\dfrac{\Sigma xy}{\sqrt{(\Sigma x^2)(\Sigma y^2)}}$
For Z-Score Data	$\dfrac{\Sigma(Z_x Z_y)}{N}$
t Formula for the r Statistic	$t = \dfrac{r - \rho}{s_r}$
r on a Sample	$r = \dfrac{\Sigma XY - (\Sigma X)(\Sigma Y)/n}{\sqrt{[\Sigma X^2 - (\Sigma X)^2/n][\Sigma Y^2 - (\Sigma Y)^2/n]}}$
Standard Error	$s_r = \sqrt{\dfrac{1 - r^2}{n - 2}}$

EXERCISES

1. For the data provided below, plot a scattergraph. What direction is the correlation?

S#	X	Y
1	12	34
2	12	37
3	14	40
4	17	41
5	18	56
6	20	55
7	21	56
8	25	60

2. For the data provided below, plot a scattergraph. What direction is the correlation?

S#	X	Y
1	11	102
2	11	100
3	12	98
4	14	90
5	14	92
6	15	84
7	17	79
8	21	65

3. Determine ρ for the data given in Exercise 1. Use the raw score formula. How much of the variance is accounted for by the association?

4. Determine ρ for the data given in Exercise 2. Use the deviation score formula. How much of the variance is accounted for by the association?

5. For the data below, determine ρ using the Z-score formula.

S#	X	Y
1	1	3
2	2	3
3	4	5
4	6	4
5	7	5
6	10	8

6. An administrator at a small junior college is curious about the relationship between student performance in chemistry (a hard science) and psychology (a soft science). She has the final exam scores for 15 students who took both courses. She finds the correlation (r) between the two exams to be 0.64. At $\alpha = .05$, use a non-directional alternate to test the hypothesis that there is no correlation in the population from which her sample was drawn.

7. Determine the correlation between the "intelligence" test scores and the "creativity" test scores given below.

	Intelligence Test	Creativity Test
Peter	122	15
Dan	102	35
Anna	135	20
Harish	110	22
Carl	140	12
Azena	130	10
Lori	128	15

CHAPTER 17

Predictive Techniques

When we know that two variables are related, we can use this information to make predictions from one variable to the other. For example, college applicants are accepted on the basis of their high school marks because it has been found that college performance is correlated with high school performance; most students who do well in high school also do well in college. Looking at this another way, we can *predict* how well a person will perform in college if we know his or her high school average. Our prediction is based on how the average student, with that particular high school average, has done in the past in college. Sometimes we will be wrong, of course, but if the correlation is very strong we will usually predict correctly.

The previous chapter discussed the straight line of best fit in a bivariate frequency distribution. Let's look at how that line is determined.

The Regression Line

The straight line of best fit is called the **regression line**. The mathematical equation that defines it is called the **regression equation**. The following scattergraph shows the regression line for a set of data.

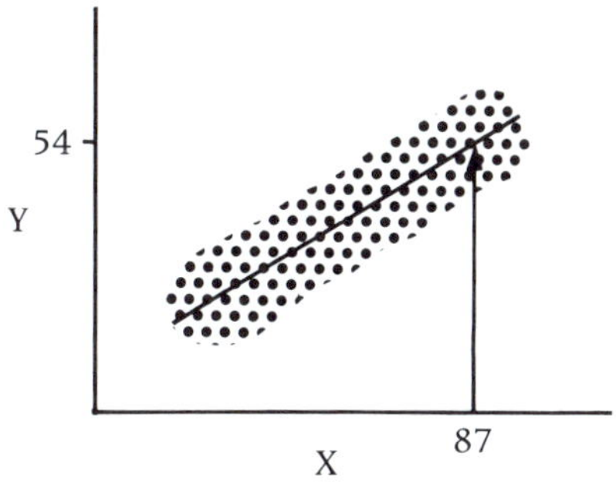

If we gather data and determine the correlation between two variables, we can use this information in the future to predict from one variable to the other. When we use the correlation between two variables to predict one from the other, we use the regression line as our prediction. We take a particular X value and predict for Y the point on the regression corresponding to that X value. We can take any value of X and predict what Y will be by noting the corresponding Y value on the line. On the scattergraph above, when X is 87, we would predict that Y is 54.

Criterion of Best Fit

> **Least Squares Criterion:** For the regression line, a mathematical criterion which specifies that the sum of the squared-distances of points from the line is a minimum.

To fit the straight line to a bivariate frequency distribution, Pearson used the **least squares criterion**. This criterion dictates that the line be laid down in such a way that the sum of the squared distances between the Y values and the line be as small as possible. In other words, *the sum of the squared distances is a minimum.* Consider the scattergraph below.

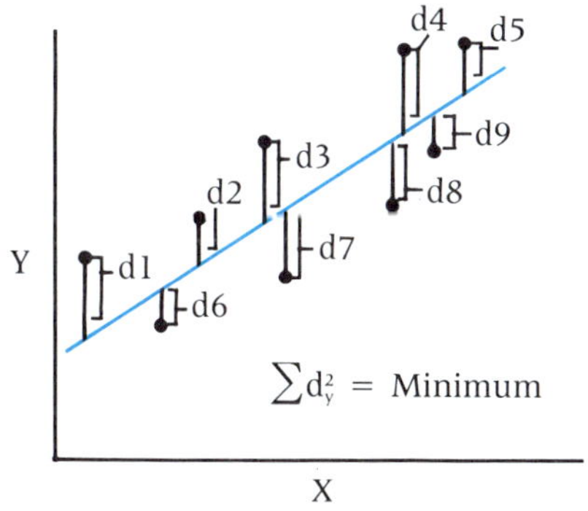

The points are the actual values of Y found for each value of X. The straight line is called the line of regression of Y on X. Note that the distances are measured in the Y dimension. The line, Y on X, is laid down so that the sum of the distances squared is as small as possible in the Y dimension. This may sound familiar. In Chapter 3, we learned one way to define a mean: the place in a distribution where the sum of the squared deviations about it is a minimum. The regression line is much like a mean.

A different line is fit to the data when we minimize the sum of the squared discrepancies in the X dimension. Consider the scattergraph below.

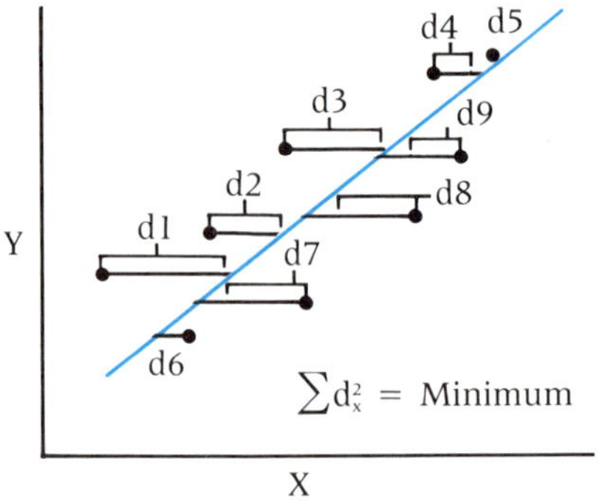

Here the distances are measured in the X dimension. This regression line of X on Y is different from Y on X. Which line is appropriate is based on how you name your variables and plot your data. We will always assume that X is known and Y is to be predicted; therefore our regression line will always by Y on X. Keep in mind, however, that two lines can always be fit to the same data.

The Regression Equation

The regression line is defined by an equation that takes the same general form as the equation for any straight line. You may recall from your high school math, the equation for a straight line is

$$Y = \underbrace{b}_{\text{slope}} X + \underbrace{a}_{\text{Y intercept}}$$

In any equation for a straight line, the slope indicates the amount of increase in Y which accompanies one unit of increase in X. As you can see below, the regression equation can be expressed in raw score, deviation score, and Z-score form. Prime notation (′) is used to indicate that Y is a predicted value. The σ_y and σ_x refer to the standard deviation of Y and the standard deviation of X, respectively.

	slope		Y intercept
Raw Score:	$Y' = \rho\left(\frac{\sigma_Y}{\sigma_X}\right)X$	$+$	$\left(-\rho\left(\frac{\sigma_Y}{\sigma_X}\right)\mu_X + \mu_Y\right)$
Deviation Score:	$y' = \rho\left(\frac{\sigma_Y}{\sigma_X}\right)x$	$+$	0
Z-score:	$Z_{Y'} = \rho Z_X$	$+$	0

In the deviation score and Z-score formula, the Y intercept is zero. The regression line passes through the ordinate at zero. The regression line always intersects at the mean of Y and the mean of X. When data are in deviation or Z-score form, the mean is zero and so the regression line must pass through zero.

In the Z-score regression equation, the value of ρ is the slope. When the data are in Z-score form, ρ indicates what portion of a *standard deviation* Y increases for *one standard deviation* increase in X.

The regression line can be determined by calculating the slope and the Y intercept of a set of data.

Calculating the Slope

When data are in raw score form the formula for the slope was

$$b = \rho\left(\frac{\sigma_Y}{\sigma_X}\right) \qquad \text{Slope of the Regression Line}$$

By substituting the raw score formula for ρ and simplifying the equation, we find that the formula for calculating the slope with raw data is

$$b = \frac{\Sigma XY - (\Sigma X)(\Sigma Y)/N}{\Sigma X^2 - (\Sigma X)^2/N} \qquad \text{Slope for Raw Data}$$

Let's use this formula to calculate the slope of the regression line for the following data.

	X	Y	XY	X^2
	3	5	15	9
	5	7	35	25
	4	4	16	16
	7	8	56	49
	6	9	54	36
	3	5	15	9
	2	3	6	4
	3	4	12	9
	4	3	12	16
	5	7	35	25
Total	42	55	256	198
Mean	4.2	5.5		

Using the formula

$$b = \frac{\Sigma XY - (\Sigma X)(\Sigma Y)/N}{\Sigma X^2 - (\Sigma X)^2/N}$$

$$= \frac{256 - (42)(55)/10}{198 - (42)^2/10}$$

$$= 25/21.6 = 1.16$$

Calculating the Y Intercept

The raw score formula for calculating the Y intercept was

$$a = -\rho\left(\frac{\sigma_Y}{\sigma_X}\right)\mu_x + \mu_y$$ Y Intercept of the Regression Line

Since $\rho\left(\frac{\sigma_Y}{\sigma_X}\right)$ is the slope, b, we can simplify the formula to

$$a = \mu_y - b\mu_x$$

Let's use the previous example to determine the Y intercept of the regression line:

$$a = 5.5 - (1.16)(4.2) = 0.63$$

The regression line passes through the Y axis at 0.63. You will recall that all regression lines pass through the point of intersection of the mean of Y and the mean of X. We have enough information to lay down the regression line for our example.

Laying Down the Regression Line

Let's lay down the regression line for the example we've been using. We know that the line crosses the Y axis at 0.63 and we know that it passes through the mean of X (i.e., 4.2) and the mean of Y (i.e., 5.5). We need only plot these two points and connect them to obtain the regression line.

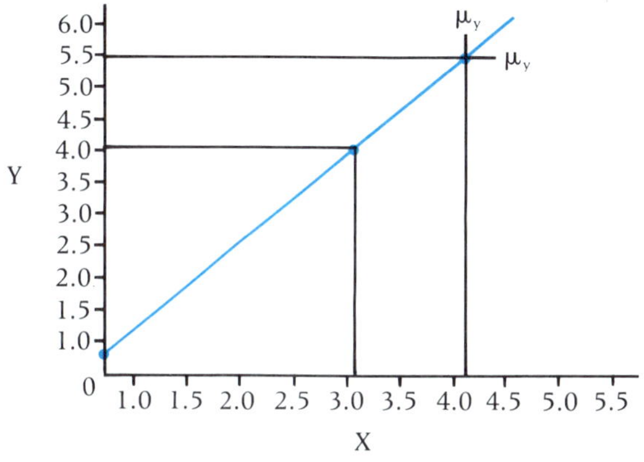

Let's verify our work by predicting the value of Y for an X value of 3. The raw score equation is

$$Y' = \rho\left(\frac{\sigma_y}{\sigma_x}\right)X - \rho\left(\frac{\sigma_y}{\sigma_x}\right)\mu_x + \mu_y$$

Since we already know the slope and Y intercept, we can simplify this equation to

$$\begin{aligned} Y' &= bX + a \\ &= 1.16(3) + 0.63 = 4.11 \end{aligned}$$

As the illustration above shows, the predicted Y value of 4.11 for an X score of 3 is, indeed, correct.

Using the Regression Equation for Prediction

Let's see how the regression equation can be used to predict Y from X. Suppose a statistics professor kept student test performance scores for many years. She found that performance on the first midterm test correlated reasonably well with the final grade students received in the course. Here are the data:

Mean score on first test: $\mu_x = 56.45$ $\quad \sigma_x = 10.25$
Mean final grade: $\mu_y = 63.12$ $\quad \sigma_y = 12.14$
The correlation between the two variables: $\rho = 0.60$

A student in her current class scored 76.0 on the first midterm exam.

Let's use the regression equation to predict the final grade this student will receive in the course.

$$Y' = \rho\left(\frac{\sigma_Y}{\sigma_X}\right) X - \rho\left(\frac{\sigma_Y}{\sigma_X}\right) \mu_X + \mu_Y$$

$$Y' = (0.60)\left(\frac{12.14}{10.25}\right)(76.0) - (0.60)\left(\frac{12.14}{10.25}\right)(56.45) + 63.12$$

$$= (0.71)(76.0) - (0.71)(56.45) + 63.12$$

$$= 54.01 - 40.11 + 63.12 = 77.02$$

We would predict a final grade of 77.02 for this student. Examine the diagram below.

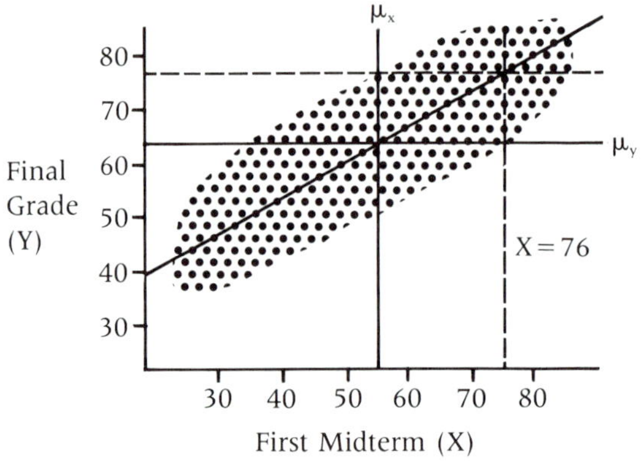

The regression equation predicts the average final grade received by students who scored 76.0 on the first midterm test. Since the correlation is not perfect there will be some error in prediction. The predicted value is the point on the line, but the actual points scatter around the regression line to some degree. How far the point vary from the line determines how much error we will make when we use the regression line as our predictor.

Error of Prediction: The Standard Error of Estimate

The regression equation states what value of Y is expected when X has a particular value. Y′ is not likely to be the actual value corresponding to X. For example, a regression equation may predict that a man who is 183 cm tall will weigh 79.5 kg. However, we would not expect a particular 183 cm tall man to weigh exactly 79.5 kg. The predicted value is an estimate of weights of men who are 183 cm tall. If the correlation between height and weight is not strong, then we can expect considerable variation in the actual values. Only when there is a perfect correlation, $\rho = \pm 1$, will the actual values precisely equal the predicted values.

The **standard error of estimate** measures the variability of the actual Y values from the predicted Y values. It is denoted as σ_{yx}. Recall that the standard deviation of Y is found by the following formula:

$$\sigma_y = \sqrt{\frac{\Sigma(Y - \mu_y)^2}{N}} \qquad \text{Standard Deviation of Y}$$

To determine the error encountered when we make predictions between two moderately correlated variables, we measure the variability of the actual Y values about the predicted values (Y′). The standard error of estimate indicates the variability of the actual Y values around the regression line. The formula for the standard error of estimate is

$$\sigma_{yx} = \sqrt{\frac{\Sigma(Y - Y')^2}{N}} \qquad \text{Standard Error of Estimate}$$

This measures the magnitude of the error of prediction. When the correlation is perfect, each $Y - Y'$ difference is zero because all the actual Y values fall on the regression line. Therefore, the standard error of estimate is zero and there is no error in prediction. When the correlation is zero, then Y′ equals the mean of Y for all X's and

$$\sigma_{yx} = \sqrt{\frac{\Sigma(Y - \mu_y)^2}{N}} = \sigma_y$$

In other words, the standard error of estimate ranges from zero when $\rho = \pm 1$ to the standard deviation of Y when $\rho = 0$.

The standard error of estimate is a standard deviation and has all the properties of one. The sum of the deviations of the Y values about the regression line is zero and the sum of the squared deviations of the Y values about the regression line is minimized.

The standard error of estimate can be calculated more easily with the following formula:

$$\sigma_{yx} = \sigma_y \sqrt{1 - \rho^2} \qquad \text{Standard Error of Estimate}$$

As we saw above, when $\rho = 0$ then $\sigma_{yx} = \sigma_y$; when $\rho = \pm 1$ then $\sigma_{yx} = 0$.

Interpreting the Correlation in Terms of Explained Variance

When X and Y are correlated, Y takes on different values depending on whether X is high or low. If we select a single X value, Y still varies to some extent unless the correlation is perfect. In other words, unless the correlation is ± 1, the actual Y values will vary around the straight line of best fit. With lower correlations, this variability increases.

The total variation in the Y distribution can be partitioned into two components:

1. the variation in Y that is associated with changes in X, and
2. the variation inherent in Y that is independent of changes in X.

Let's take a single Y value and look at how its variation can be partitioned into two parts.

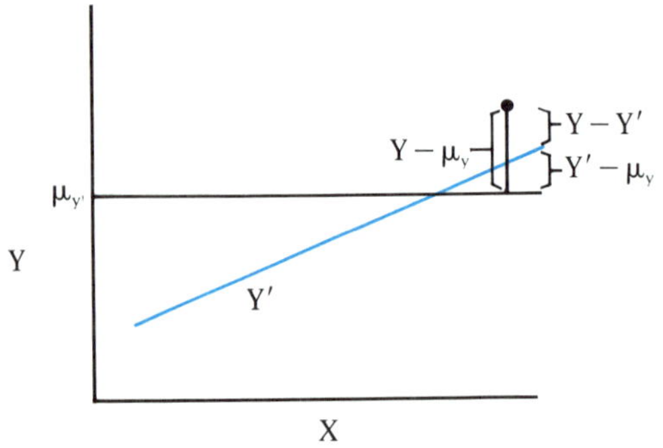

Remember that the total variation in Y is based on the deviations of the Y values from the mean of the distribution. If all the deviations were squared, summed and divided by N, we would have the Y variance.

Look at the graph above. The Y value differs from the mean of the Y distribution. That discrepancy can be partitioned into (1) the difference between the Y value and the point on the straight line of best fit (i.e., Y′), and (2) the difference between the line (Y′) and the mean of the distribution.

If we square all the differences, sum the squares, and then divide each by N, we would have the following:

$$\frac{\Sigma(Y - \mu_y)^2}{N} = \frac{\Sigma(Y - Y')^2}{N} + \frac{\Sigma(Y' - \mu_y)^2}{N}$$

$$\sigma_y^{\ 2} = \sigma_{yx}^{\ 2} + \sigma_{y'}^{\ 2}$$

$$\text{Total Y variance} = \begin{pmatrix}\text{Variance in Y}\\ \text{independent of}\\ \text{changes in X}\end{pmatrix} + \begin{pmatrix}\text{Variance in Y}\\ \text{associated with}\\ \text{changes in X}\end{pmatrix}$$

where $\sigma_y^{\ 2}$ is the total variance in the Y distribution; $\sigma_{yx}^{\ 2}$ is the variance of the actual Y values around the straight line (Y′); and $\sigma_{y'}^{\ 2}$ is the variance of the points on the line of best fit around the mean of the Y distribution.

When the correlation is ± 1, all the Y values fall exactly on the straight line of best fit, and there is no variation of the Y values from Y′. $\sigma_{yx}^{\ 2}$ will be zero; all the variation in Y is due to changes in X. When the correlation is 0, none of the variation in Y is due to changes in X; $\sigma_{y'}^{\ 2}$ will be zero.

The correlation coefficient can be interpreted in terms of the *proportion of the total* Y *variance which is associated with changes in* X. This proportion is called the **coefficient of determination**. It is determined by

$$\frac{\text{Variance in Y associated with changes in X}}{\text{Total Y variance}} = \frac{\sigma_{y'}^{\ 2}}{\sigma_y^{\ 2}} = \rho^2$$

Recall that the coefficient of determination, ρ^2, indicates the strength of the relationship between the two variables. It estimates how much of the total Y variation is due to the correlation Y has with X.

Think of the coefficient of determination in terms of how much of the variance in Y can be *explained* and how much is left unaccounted for. Many people find this a more helpful way to think about correlations. For example, if you have a correlation of 0.50, you know that 25% of the total variability of Y is due to the correlation and the rest is unexplained.

Regression on the Mean

When a value of Y is predicted from a given value of X, Y′ can never be further from its mean than X is from its mean. Unless the correlation is perfect, Y′ will be closer to its mean than X is to its mean. This is called **regression on the mean**. It can lead us into some peculiar situations.

Consider the correlation between IQ of parents and their offspring. It's about 0.5. Let's predict the IQ of children whose parents' IQ are 2 standard deviations below the mean. We will use our Z-score formula.

$$\begin{aligned} Z_{y'} &= \rho Z_x + 0 \\ &= 0.5(-2) + 0 = -1 \end{aligned}$$

For parents whose IQ scores are 2 standard deviations below the mean, we will predict their offspring to be only 1 standard deviation below the mean. Similarly, for parents whose scores are 2 standard deviations above the mean, we will predict their offspring to be 1 standard deviation above the mean. As you can see, the predicted values regress toward the mean.

Examine the following diagram.

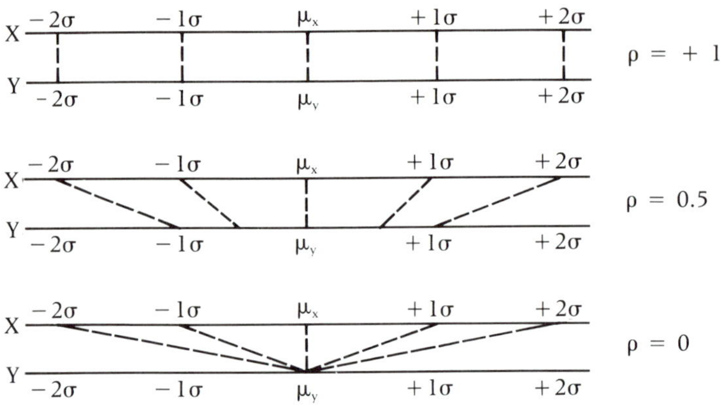

The diagram shows the Z-scores of Y we would predict for various Z-scores of X, when ρ is 0, 0.5, and 1. When $\rho = 0$, we predict that the Z score of Y will be the mean (which in any Z distribution is 0). When $\rho = 1$, the Z-score of Y will be the same as the Z-score of X. But when ρ is 0.5, the predicted value of Y will tend toward the mean.

Regression on the mean occurs because the predicted Y score cannot be further away from the mean of its distribution than its X counterpart. We must always predict that Y will be no further away from its mean than X is from its mean.

SUMMARY OF TERMS AND FORMULAS

When the **correlation** between two variables is known, we may use knowledge of performance on the first variable, to **predict** performance on the second variable. A **regression line** is fitted to the correlational data and the point that lies on the line is the predicted value of Y for a given X value. The amount of **predictive error** is determined by the strength of the correlation. If the correlation is not perfect, some predictive error will occur. The measure of predicted error is called the **standard error of estimate**.

The **coefficient of determination**, ρ^2, can be interpreted in terms of the proportion of the total variance in Y that can be explained or accounted for by the relationship of Y to X.

Regression on the mean occurs in prediction whenever the correlation between X and Y is less than 1. The predicted value will tend to be closer to its mean than was the value used in the prediction.

Regression Equation for Predicting Y from X

Raw Score: $$Y' = \rho\left(\frac{\sigma_y}{\sigma_x}\right)X - \rho\left(\frac{\sigma_y}{\sigma_x}\right)\mu_x + \mu_y$$

Deviation Score: $$y' = \rho\left(\frac{\sigma_y}{\sigma_x}\right)x$$

Z-score: $$Z_{y'} = \rho Z_x$$

Slope of the Regression Line

$$b = \rho\left(\frac{\sigma_y}{\sigma_x}\right)$$

Slope for Raw Data $$b = \frac{\Sigma XY - (\Sigma X)(\Sigma Y)/N}{\Sigma X^2 - (\Sigma X)^2/N}$$

Y Intercept of the Regression Line

$$a = -\rho\left(\frac{\sigma_y}{\sigma_x}\right)\mu_x + \mu_y$$

Standard Error of Estimate

$$\sigma_{yx} = \sqrt{\frac{\Sigma(Y - Y')^2}{N}}$$

$$\sigma_{yx} = \sigma_y\sqrt{1 - \rho^2}$$

EXERCISES

A professor collected the following data on number of hours per week students spend studying and their scores on tests.

#hrs/wk	Test Scores
8	75
8	50
2	50
4	45
4	65
9	60
10	80
10	95
2	35
5	50
7	65
4	60
2	50
1	50
7	70

1. Calculate the slope for the above data (use the raw score equation).

2. Calculate the Y intercept for the above data.

3. Plot the regression line. Be precise.

4. If Russell tells you he studies 7.5 hours/week, circle the test score on the graph from question 3 that you would predict for him.

5. Verify the predicted value from question 4 with your equation.

6. $\rho = 0.60$, $\sigma_y = 4.2$, $\sigma_x = 3.0$, $\mu_x = 8.6$, $\mu_y = 12.0$
Determine Y′ for an X value of:

a) 2
b) 8
c) 5

7. If $\rho = -0.60$, what Z-score in Y would be predicted for:

a) a score 1/2 a standard deviation below the mean in X
b) $Z_x = 1.5$
c) a score equal to the mean of X

CHAPTER 18

Toolbox

This chapter consists of two parts. The first section provides most of the computational formulas used in the text. This provides you with a handy reference. The second section contains all the test summaries found in the text. You are encouraged to refer to the appropriate chapter before beginning any analysis.

Toolbox of Computational Formulas

Analysis of Variance

One-Way ANOVA

Sums of Squares

Total $$SS_{TOT} = \Sigma X_{tot}^2 - \frac{(\Sigma X_{tot})^2}{n_{tot}}$$

Between Groups $$SS_{BG} = \frac{(\Sigma X_1)^2}{n_1} + \frac{(\Sigma X_2)^2}{n_2} + \ldots + \frac{(\Sigma X_k)^2}{n_k} - \frac{(\Sigma X_{tot})^2}{n_{tot}}$$

Within Groups $$SS_{WG} = SS_{TOT} - SS_{BG}$$

Mean Squares

Between Groups $$MS_{BG} = \frac{SS_{BG}}{k - 1}$$

Within Groups $$MS_{WG} = \frac{SS_{WG}}{n_{tot} - k}$$

F Ratio $$F = MS_{BG}/MS_{WG}$$

Two-Way ANOVA

Sums of Squares

Total $$SS_{TOT} = \Sigma X_{tot}^2 - \frac{(\Sigma X_{tot})^2}{n_{tot}}$$

Between Groups $$SS_{BG} = \frac{(\Sigma A_1B_1)^2 + (\Sigma A_1B_2)^2 + \ldots (\Sigma A_aB_b)^2}{n} - \frac{(\Sigma X_{tot})^2}{n_{tot}}$$

A $$SS_A = \frac{(\Sigma A_1)^2 + (\Sigma A_2)^2 + \ldots + (\Sigma A_a)^2}{bn} - \frac{(\Sigma X_{tot})^2}{n_{tot}}$$

B $$SS_B = \frac{(\Sigma B_1)^2 + (\Sigma B_2)^2 + \ldots + (\Sigma B_b)^2}{an} - \frac{(\Sigma X_{tot})^2}{n_{tot}}$$

AXB $$SS_{AXB} = SS_{BG} - SS_A - SS_B$$

Within Groups $$SS_{WG} = SS_{TOT} - SS_{BG}$$

Mean Squares

A $$MS_A = \frac{SS_A}{a - 1}$$

B $$MS_B = \frac{SS_B}{b - 1}$$

AXB $$MS_{AxB} = \frac{SS_{AxB}}{(a - 1)(b - 1)}$$

WG $$MS_{WG} = \frac{SS_{WG}}{n_{tot} - k}$$

F Ratios

$$MS_A/MS_{WG}$$

$$MS_B/MS_{WG}$$

$$MS_{AXB}/MS_{WG}$$

One-Way ANOVA with Repeated Measures

Sums of Squares

Total
$$SS_{TOT} = \Sigma X_{tot}^2 - \frac{(\Sigma X_{tot})^2}{kn}$$

Between Subject
$$SS_S = \frac{(\Sigma S_1)^2 + (\Sigma S_2)^2 + \ldots + (\Sigma S_n)^2}{k} - \frac{(\Sigma X_{tot})^2}{kn}$$

Within Subject
$$SS_{WS} = SS_{TOT} - SS_S$$

Treatment
$$SS_T = \frac{(\Sigma T_1)^2 + (\Sigma T_2)^2 + \ldots + (\Sigma T_k)^2}{n} - \frac{(\Sigma X_{tot})^2}{kn}$$

Subject by Treatment
$$SS_{SXT} = SS_{WS} - SS_T$$

Mean Squares

Treatment
$$MS_T = \frac{SS_T}{k - 1}$$

Subject by Treatment
$$MS_{SXT} = \frac{SS_{SXT}}{(n - 1)(k - 1)}$$

F-Ratio

$$F = \frac{MS_T}{MS_{SXT}}$$

Two-Way ANOVA with Repeated Measures

Sums of Squares

Total
$$SS_{TOT} = \Sigma X_{tot}^2 - \frac{(\Sigma X_{tot})^2}{abn}$$

Between Subjects
$$SS_S = \frac{(\Sigma S_1)^2 + (\Sigma S_2)^2 + \ldots + (\Sigma S_{an})^2}{b} - \frac{(\Sigma X_{tot})^2}{abn}$$

A
$$SS_A = \frac{(\Sigma A_1)^2 + (\Sigma A_2)^2 + \ldots + (\Sigma A_a)^2}{bn} - \frac{(\Sigma X_{tot})^2}{abn}$$

Subjects within Groups $SS_{S(gps)} = SS_S - SS_A$

B $$SS_B = \frac{(\Sigma B_1)^2 + (\Sigma B_2)^2 + \ldots + (\Sigma B_b)^2}{an} - \frac{(\Sigma X_{tot})^2}{abn}$$

AXB $$SS_{AXB} = \frac{(\Sigma A_1B_1)^2 + (\Sigma A_1B_2)^2 + \ldots + (\Sigma A_aB_b)^2}{n} - \frac{(\Sigma X_{tot})^2}{abn} - SS_A - SS_B$$

Subjects within Groups XB $SS_{S(gps)XB} = SS_{WS} - SS_B - SS_{AXB}$

Mean Squares

A $$MS_A = \frac{SS_A}{a - 1}$$

Subjects within Groups $$MS_{S(gps)} = \frac{SS_{S(gps)}}{a(n - 1)}$$

B $$MS_B = \frac{SS_B}{b - 1}$$

AXB $$MS_{AXB} = \frac{SS_{AXB}}{(a - 1)(b - 1)}$$

Subjects within Groups XB $$MS_{S(gps)XB} = \frac{SS_{S(gps)XB}}{a(n - 1)(b - 1)}$$

F-Ratios

$$F_A = \frac{MS_A}{MS_{S(gps)}}$$

$$F_B = \frac{MS_B}{MS_{S(gps)XB}}$$

$$F_{AXB} = \frac{MS_{AXB}}{MS_{S(gps)XB}}$$

Chi-Square Formula

$$\chi^2 = \Sigma \frac{(0 - E)^2}{E}$$

Combinations

$${}_nC_r = \frac{n!}{(n - r)!r!}$$

Confidence Intervals

Difference: $\mu_1 - \mu_2 = (\overline{X}_1 - \overline{X}_2) \pm Z(\sigma_{\overline{X}1-\overline{X}2})$

Mean: $\mu = \overline{X} \pm Z(\sigma_{\overline{X}})$

Proportion: $P = p \pm Z(\sigma_p)$

When σ is estimated with s_i

Difference-Independent: $\mu_1 - \mu_2 = (\overline{X}_1 - \overline{X}_2) \pm t(s_{\overline{X}1-\overline{X}2})$

Difference-Dependent: $\mu_{\overline{D}} = \overline{D} \pm t(s_{\overline{D}})$

Mean: $\mu = \overline{X} \pm t(s_{\overline{X}})$

Deviation Score

$x = X - \mu$

Kruskal-Wallis H

$$H = \frac{12}{n_{tot}(n_{tot}+1)}\left[\frac{(\Sigma R_1)^2}{n_1} + \frac{(\Sigma R_2)^2}{n_2} + \ldots + \frac{(\Sigma R_k)^2}{n_k}\right] - 3(n_{tot}+1)$$

Mann-Whitney U

$$U_1 = n_1 n_2 + \frac{n_1(n_1+1)}{2} - \Sigma R_1$$

$$U_2 = n_1 n_2 + \frac{n_2(n_2+1)}{2} - \Sigma R_2$$

Mean

Population

Raw Data: $\mu = \dfrac{\Sigma X}{N}$

Data in a Frequency Distribution: $\mu = \dfrac{\Sigma fX}{N}$

Mean for Combined Subgroups: $\mu_c = \dfrac{\Sigma X + \Sigma Y}{N_x + N_y}$

when sums are known: $\mu_c = \dfrac{\Sigma X + \Sigma Y}{N_x + N_y}$

when means are known: $\mu_c = \dfrac{N_x\mu_x + N_y\mu_y}{N_x + N_y}$

when $N_x = N_y$: $\mu_c = \dfrac{\mu_x + \mu_y}{2}$

Sample

Raw Data: $\bar{X} = \frac{\Sigma X}{n}$

Data in a Frequency Distribution: $\bar{X} = \frac{\Sigma fX}{n}$

Median

$$Mdn = L + \frac{\left[N\left(\frac{50}{100}\right) - f_b\right]i}{f_w}$$

Pearson's Coefficient of Correlation

Raw Data:

$$\rho = \frac{\Sigma XY - (\Sigma X)(\Sigma Y)/N}{\sqrt{[\Sigma X^2 - (\Sigma X)^2/N][\Sigma Y^2 - (\Sigma Y)^2/N]}}$$

For Deviation Data:

$$\rho = \frac{\Sigma \chi\gamma}{\sqrt{(\Sigma \chi^2)(\Sigma \gamma^2)}}$$

For Z-Score Data:

$$\rho = \frac{\Sigma(Z_x Z_y)}{N}$$

When estimating ρ with r:

$$r = \frac{\Sigma XY - (\Sigma X)(\Sigma Y)/n}{\sqrt{[\Sigma X^2 - (\Sigma X)^2/n][\Sigma Y^2 - (\Sigma Y)^2/n]}}$$

Percentile

$$P_{PR} = L + \frac{\left[N\left(\frac{PR}{100}\right) - f_b\right]i}{f_w}$$

Percentile Rank

$$PR_X = \frac{\left[f_w \frac{(X - L)}{i} + f_b\right]100}{N}$$

Permutations

$${}_nP_r = \frac{n!}{(n - r)!}$$

Probability

Binomial Probability

$${}_nC_r\, p^r q^{n-r} = \frac{n!}{(n - r)!r!} p^r q^{n-r}$$

Compound

p (A and B)

Dependent events: $p(A) \cdot p(B/A)$

Independent events: $p(A) \cdot p(B)$

p (A or B)

Not mutually exclusive: $p(A) + p(B) - p(A \text{ and } B)$

Mutually exclusive: $p(A) + p(B)$

Conditional $$p(B/A) = \frac{\#B/A \text{ has occurred}}{\#O/A \text{ has occurred}}$$

Simple $$p(A) = \frac{\#A}{\#O}$$ Given equally likely events

---

Range

$H - L + 1$

Regression Formulas

Raw Score: $$Y' = \rho\left(\frac{\sigma_y}{\sigma_x}\right)X - \rho\left(\frac{\sigma_y}{\sigma_x}\right)\mu_x + \mu_y$$

Deviation Score: $$y' = \rho\left(\frac{\sigma_y}{\sigma_x}\right)x$$

Z-Score: $$Z_{y'} = \rho Z_x$$

Slope of the Regression Line: $$b = \rho\left(\frac{\sigma_y}{\sigma_x}\right)$$

Slope for Raw Data: $$b = \frac{\Sigma XY - (\Sigma X)(\Sigma Y)/N}{\Sigma X^2 - (\Sigma X)^2/N}$$

Y intercept of the Regression Line: $$a = -\rho\left(\frac{\sigma_y}{\sigma_x}\right)\mu_x + \mu_y$$

Rank-Order Correlation Coefficient

$$\text{rho} = 1 - \frac{6\Sigma d^2}{n(n^2 - 1)}$$

Scheffé

F′ Statistic:	$F' = C/s_c$
Scheffé's Critical F:	$F_s = \sqrt{(k - 1)F_{crit}}$
Comparison:	$C = c_1\overline{X}_1 + c_2\overline{X}_2 + \ldots + C_k\overline{X}_k$

Standard Deviation

Population

Raw Data: $$\sigma = \sqrt{\frac{\Sigma X^2 - (\Sigma X)^2/N}{N}}$$

Data in Frequency Distribution: $$\sigma = \sqrt{\frac{\Sigma fX^2 - (\Sigma fX)^2/N}{N}}$$

Estimate of σ: $$s = \sqrt{\frac{\Sigma X^2 - (\Sigma X)^2/n}{n - 1}}$$

Standard Errors: for Z-tests

Difference: $$\sigma_{\overline{x}1-\overline{x}2} = \sqrt{\frac{\Sigma x_1^2 + \Sigma x_2^2}{n_1 + n_2}\left(\frac{1}{n_1} + \frac{1}{n_2}\right)}$$

$$\sigma_{\overline{x}1-\overline{x}2} = \sqrt{\frac{\Sigma x_1^2 + \Sigma x_2^2}{n(n)}} \quad \text{if } n_1 = n_2$$

Mean $$\sigma_{\overline{x}} = \sigma/\sqrt{n}$$

Proportion $$\sigma_p = \sqrt{PQ/n}$$

Standard Errors: Estimates for t-tests

Difference

Independent Samples $$s_{\overline{x}1-\overline{x}1} = \sqrt{s_{\overline{x}1}^2 + s_{\overline{x}2}^2}$$

$$s_{\overline{x}1-\overline{x}2} = \sqrt{\frac{\Sigma x_1^2 + \Sigma x_2^2}{n_1 + n_2 - 2}\left(\frac{1}{n_1} + \frac{1}{n_2}\right)} \quad \text{if } n_1 \neq n_2$$

$$s_{\overline{x}1-\overline{x}2} = \sqrt{\frac{\Sigma x^2 + \Sigma x_2^2}{n(n - 1)}} \quad \text{if } n_1 = n_2$$

Dependent Samples $$s_{\overline{D}} = \sqrt{\frac{\Sigma D^2 - (\Sigma D)^2/n}{n(n - 1)}}$$

Mean $$s_{\bar{x}} = s/\sqrt{n}$$

$$= \sqrt{\frac{\Sigma X^2 - (\Sigma X)^2/n}{n(n - 1)}}$$

Standard Error for a Comparison

$$s_c = \sqrt{MS_{error}\left(\frac{c_1^{\,2}}{n_1} + \frac{c_2^{\,2}}{n_2} + \ldots + \frac{c_k^{\,2}}{n_k}\right)}$$

Standard Error of Estimate

$$\sigma_{yx} = \sqrt{\frac{\Sigma(Y - Y')^2}{N}}$$

$$\sigma_{yx} = \sigma_y\sqrt{1 - \rho^2}$$

Standard Error for Pearson's r

$$s_r = \sqrt{\frac{1 - r^2}{n - 2}}$$

---

t-Ratios

Mean: $$t = \frac{\bar{X} - \mu_{\bar{x}}}{s_{\bar{x}}}$$

Difference: Independent Means $$t = \frac{\bar{X}_1 - \bar{X}_2}{s_{\bar{x}1 - \bar{x}2}}$$

Difference: Dependent Means $$t = \frac{\bar{D} - \mu_{\bar{D}}}{s_{\bar{D}}}$$

r $$t = \frac{r - \rho}{s_r}$$

Tukey

$$HSD = q(\alpha, df_{error}, k)\sqrt{MS_{error}/n}$$

Variance

Population

Raw Data $$\sigma^2 = \frac{\Sigma X^2 - (\Sigma X)^2/N}{N}$$

Data in Frequency Distribution $\sigma^2 = \dfrac{\Sigma fX^2 - (\Sigma fX)^2/N}{N}$

Pooled Variance for Three Combined Subgroups

$$\sigma_c^2 = \frac{N_w\sigma_w{}^2 + N_x\sigma_x{}^2 + N_y\sigma_y{}^2 + N_w(\mu_w - \mu_c)^2 + N_x(\mu_x - \mu_c)^2 + N_y(\mu_y - \mu_c)^2}{N_w + N_x + N_y}$$

Estimate of σ^2 $s^2 = \dfrac{\Sigma X^2 - (\Sigma X)^2/n}{n - 1}$

Wilcoxon $Z = \dfrac{T - \mu_T}{\sigma_T}$

Z-Ratios

Difference $Z = \dfrac{\overline{X}_1 - \overline{X}_2}{\sigma_{\overline{x}1-\overline{x}2}}$

Mean $Z = \dfrac{\overline{X} - \mu_{\overline{x}}}{\sigma_{\overline{x}}}$

Proportion $Z = \dfrac{p - P}{\sigma_p}$

Score $Z = \dfrac{X - \mu}{\sigma}$

Mann-Whitney U $Z = \dfrac{U - n_1n_2/2}{\sqrt{\dfrac{n_1n_2(n_1 + n_2 + 1}{12}}}$

Wilcoxon T $Z = \dfrac{T - n(n + 1)/4}{\sqrt{\dfrac{n(n + 1)(2n + 1)}{24}}}$

Toolbox of Test Summaries

Summary of One-Way ANOVA

Hypotheses

H_o: $\mu_1 = \mu_2 = \ldots = \mu_k$
H_a: H_o is false

Assumptions

1. Subjects are randomly selected and independently assigned to groups.
2. Population distributions are normal.
3. Population variances are homogeneous.

Decision Rules

$df_{bg} = k - 1$ $df_{wg} = n_{tot} - k$
If $F_{obt} \geqslant F_{crit}$, reject the H_o
If $F_{obt} < F_{crit}$, do not reject H_o

Formula

$$F = \frac{MS_{BG}}{MS_{WG}}$$

Summary of Two-Way ANOVA

Hypotheses

H_o: No main effects and no interaction
H_a: H_o is false

Assumptions

1. Subjects are randomly selected and independently assigned to groups.
2. Population distributions are normal.
3. Population variances are homogeneous.

Decision Rules

$df_a = a - 1$ $df_b = b - 1$ $df_{axb} = (a - 1)(b - 1)$
If $F_{obt} \geq F_{crit}$, reject the H_o
If $F_{obt} < F_{crit}$, do not reject H_o

$$F_A = \frac{MS_A}{MS_{WG}}$$

$$F_B = \frac{MS_B}{MS_{WG}}$$

$$F_{AXB} = \frac{MS_{AXB}}{MS_{WG}}$$

Summary of One-Way ANOVA with Repeated Measures

Hypotheses

H_o: $\mu_1 = \mu_2 = \ldots = \mu_k$
H_a: H_o is false

Assumptions

1. Subjects are randomly selected.
2. Population distributions are normal.
3. Population variances are homogeneous.
4. Population covariances are equal.

Decision Rules

$df_t = k - 1 \quad df_{sxt} = (n - 1)(k - 1)$
If $F_{obt} \geqslant F_{crit}$, reject the H_o
If $F_{obt} < F_{crit}$, do not reject H_o

Formula

$$F = \frac{MS_T}{MS_{SXT}}$$

Summary of Two-Way ANOVA with Repeated Measures

Hypotheses

H_o: No main effects and no interaction

H_a: H_o is false

Assumptions

1. Subjects are randomly selected with repeated measures on factor B.
2. Population distributions are normal.
3. Population variances are homogeneous.
4. Population covariances are equal.

Decision Rules

$df_a = a - 1 \quad df_b = b - 1 \quad df_{axb} = (a - 1)(b - 1)$

$df_{s(gps)} = a(n - 1) \quad df_{s(gps)xb} = a(n - 1)(b - 1)$

If $F_{obt} \geq F_{crit}$, reject the H_o

If $F_{obt} < F_{crit}$, do not reject H_o

Formulas

$$F_A = \frac{MS_A}{MS_{S(gps)}}$$

$$F_B = \frac{MS_B}{MS_{S(gps)XB}}$$

$$F_{AXB} = \frac{MS_{AXB}}{MS_{S(gps)XB}}$$

Summary of the Chi-Square Test for Goodness of Fit

Hypotheses

H_o: O's = E's
H_a: O's ≠ E's

Assumptions

1. Subjects are randomly selected.
2. Categories are mutually exclusive.

Decision Rules

df = # categories − 1

If $\chi^2_{obt} \geq \chi^2_{crit}$, reject H_o

If $\chi^2_{obt} < \chi^2_{crit}$, do not reject H_o

Formula

$$\chi^2 = \Sigma\frac{(O - E)^2}{E}$$

Summary of the Chi-Square Test for Independence

Hypotheses

H_o: The variables are independent.
H_a: The variables are dependent.

Assumptions

1. Subjects are randomly selected.
2. Observations have been classified simultaneously on two independent categories.

Decision Rules

df = (# rows − 1)(# columns − 1)

If $\chi^2_{obt} \geq \chi^2_{crit}$, reject H_o

If $\chi^2_{obt} < \chi^2_{crit}$, do not reject H_o

Formula

$$\chi^2 = \Sigma \frac{(O - E)^2}{E}$$

Summary of the Mann-Whitney U Test

Hypotheses

H_o: Populations are identical
H_a: Populations are not identical

Assumptions

1. Subjects are randomly selected and independently assigned to groups.
2. Measurement scale is at least ordinal.

Decision Rules

If $U_{obt} \leq U_{crit}$, reject the H_o
If $U_{obt} > U_{crit}$, do not reject the H_o

Formula

$$U_1 = n_1n_2 + \frac{n_1(n_2 + 1)}{2} - \Sigma R_1$$

$$U_2 = n_1n_2 + \frac{n_2(n_2 + 1)}{2} - \Sigma R_2$$

U_{obt} is smaller of U_1 and U_2

Summary of the Pearson Correlation Test

Hypotheses

H_o: $\rho = 0$
H_a: $\rho \neq 0, \rho < 0, \rho > 0$

Assumptions

1. Subjects are randomly selected.
2. Both populations are normally distributed.

Decision Rules

If $t_{obt} \geq t_{crit}$, reject H_o
If $t_{obt} < t_{crit}$, do not reject H_o

Formula

$$t = \frac{r - \rho}{\sqrt{\frac{1 - r^2}{n - 2}}}$$

Summary of the Rank-Order Correlation Test

Hypotheses

H_o: Rho = 0

H_a: Rho ≠ 0, Rho < 0, Rho > 0

Assumptions

1. Subjects are randomly selected.
2. Observations are rank-ordered.

Decision Rules

n = number of pairs of ranks

If $rho_{obt} \geqslant rho_{crit}$, reject H_o

If $rho_{obt} < rho_{crit}$, do not reject H_o

$$rho = 1 - \frac{6\Sigma d^2}{n(n^2 - 1)}$$

Summary of the Scheffé Test

Hypotheses

H_o: each comparisons is zero
H_a: each comparisons is not zero

Assumptions

The outcome of the ANOVA was significant.

Decision Rules

If $F' \geqslant F_s$ reject the H_o
If $F' < F_s$ do not reject H_o

Formula

$F' = C/s_c$

$F_s = \sqrt{(k - 1)(F_{crit})}$

Summary of t-Test for a Single Mean

Hypotheses

H_o: μ = specified value
H_a: $\mu \neq, >, <$ specified value

Assumptions

1. Subjects are randomly selected.
2. Population distribution is normal.

Decision Rules

$df = n - 1$
If $t_{obt} \geq t_{crit}$, reject the H_o
If $t_{obt} < t_{crit}$, do not reject H_o

Formula

$$t = \frac{\overline{X} - \mu}{\sqrt{\dfrac{\Sigma X^2 - (\Sigma X)^2/n}{n(n-1)}}}$$

Summary of t-Test for Dependent Means

Hypotheses

H_o: $\mu_1 = \mu_2$

H_a: $\mu_1 \neq \mu_2$, $\mu_1 < \mu_2$, $\mu_1 > \mu_2$

Assumptions

1. Subjects are randomly selected.
2. Population distributions are normal.
3. Population variances are homogeneous.
4. Repeated measures or matched subjects are used.

Decision Rules

$df = n_{pairs} - 1$

If $t_{obt} \geq t_{crit}$, reject H_o.

If $t_{obt} < t_{crit}$, do not reject H_o.

Formula

$$t = \frac{\overline{D}}{\sqrt{\dfrac{\Sigma D^2 - (\Sigma D)^2/n}{n(n-1)}}}$$

Summary of t-Test for Independent Means

Hypotheses

H_o: $P = \mu_1 = \mu_2$
H_a: $\mu_1 \neq \mu_2$, $\mu_1 < \mu_2$, $\mu_1 > \mu_2$

Assumptions

1. Subjects are randomly selected and independently assigned to groups.
2. Population variances are homogeneous.
3. Population distributions are normal.

Decision Rules

$df = n_1 + n_2 - 2$
If $t_{obt} \geqslant t_{crit}$, reject H_o
If $t_{obt} < t_{crit}$, do not reject H_o

Formula

$$t = \frac{\overline{X}_1 - \overline{X}_2}{\sqrt{\left(\frac{\Sigma x_1^2 + \Sigma x_2^2}{n_1 + n_2 - 2}\right)\left(\frac{1}{n_1} + \frac{1}{n_2}\right)}}$$

Summary of the Tukey Tes

Hypotheses

H_o: no difference between population means
H_a: H_o is false

Assumptions

The outcome of the ANOVA was significant.

Decision Rules

Any mean difference $\geq$ HSD, reject the H_o

Formula

$$HSD = q(\alpha, df_{error}, k)\sqrt{MS_{error}/n}$$

Summary of the Wilcoxon Test

Hypotheses

H_o: Populations are identical
H_a: Populations are not identical

Assumptions

1. Subjects are randomly selected.
2. Same or matched subjects.
3. Measurement scale is at least ordinal.

Decision Rules

If $T_{obt} \leq T_{crit}$, reject the H_o
If $T_{obt} > T_{crit}$, do not reject the H_o

Formula

N = number of pairs with non-zero differences
T is the sum of the absolute ranks with the less frequently appearing sign

Summary of Z-Test for Independent Means

Hypotheses

H_o: $\mu_1 = \mu_2$
H_a: $\mu_1 \neq \mu_2$, $\mu_1 < \mu_2$, $\mu_1 > \mu_2$

Assumptions

1. Subjects are randomly selected and independently assigned to groups.
2. Population distributions are normal.
3. Population standard deviations are known.

Decision Rules

If $Z_{obt} \geq Z_{crit}$, reject H_o
If $Z_{obt} < Z_{crit}$, do not reject H_o

Formula

$$Z = \frac{\overline{X}_1 - \overline{X}_2}{\sqrt{\left(\frac{\Sigma x_1^2 + \Sigma x_2^2}{n_1 + n_2}\right)\left(\frac{1}{n_1} + \frac{1}{n_2}\right)}}$$

Summary of Z-Test for a Proportion

Hypotheses

H_o: P = specified value
H_a: P ≠, <, > specified value

Assumptions

1. Subjects are randomly selected.
2. Sampling distribution of the statistic is normal.
3. Observations are dichotomous.

Decision Rules

If $Z_{obt} \geq Z_{crit}$, reject the H_o
If $Z_{obt} < Z_{crit}$, do not reject H_o

Formula

$$Z = \frac{p - P}{\sqrt{PQ/n}}$$

Summary of Z-Test for a Single Mean

Hypotheses

H_o: μ = specified value
H_a: μ $\neq$, $<$, $>$ specified value

Assumptions

1. Subjects are randomly selected.
2. Population distribution is normal.
3. Population standard deviation is known.

Decision Rules

If $Z_{obt} \geq Z_{crit}$, reject the H_o
If $Z_{obt} < Z_{crit}$, do not reject H_o

Formula

$$Z = \frac{\overline{X} - \mu}{\sigma/\sqrt{n}}$$

Appendix

Appendix A
Answers to Exercises

Chapter 1

1. **a)** continuous **b)** discrete **c)** continuous **d)** continuous **e)** discrete **f)** discrete **g)** continuous

2. IV-brand of pain reliever
DV-rating of effectiveness

3. **a)** N = 11 **b)** 4 **c)** 6 **d)** 2

4. **a)** ordinal, discrete **b)** ratio, discrete **c)** nominal, discrete **d)** nominal, discrete

5. IV-list type
DV-time to learn

Chapter 2

1. Simple Frequency Distribution

X	f
25	1
24	0
23	0
22	0
21	0
20	0
19	0
18	1
17	0
16	0
15	3
14	1
13	1
12	1
11	1
10	1
9	1
8	2
7	1
6	1

2.

Interval	f
46-50	1
41-45	2
36-40	2
31-35	0
26-30	0
21-25	1
16-20	0
11-15	2
6-10	5
1-5	1

3.

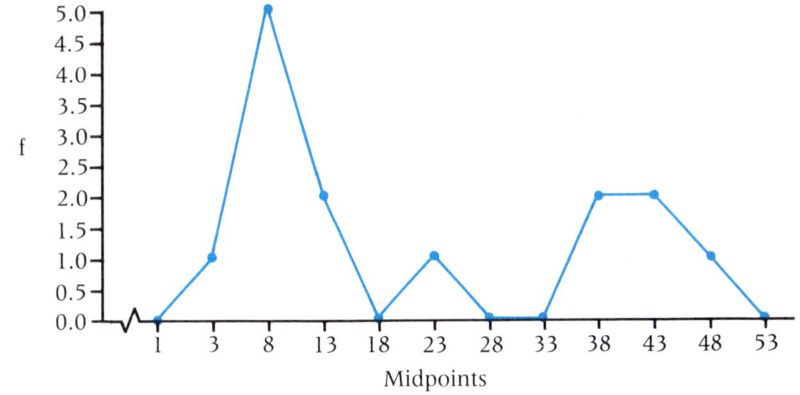
5.0
4.5
4.0
3.5
3.0
2.5
2.0
1.5
1.0
0.5
0.0
f
1 3 8 13 18 23 28 33 38 43 48 53
Midpoints

4.

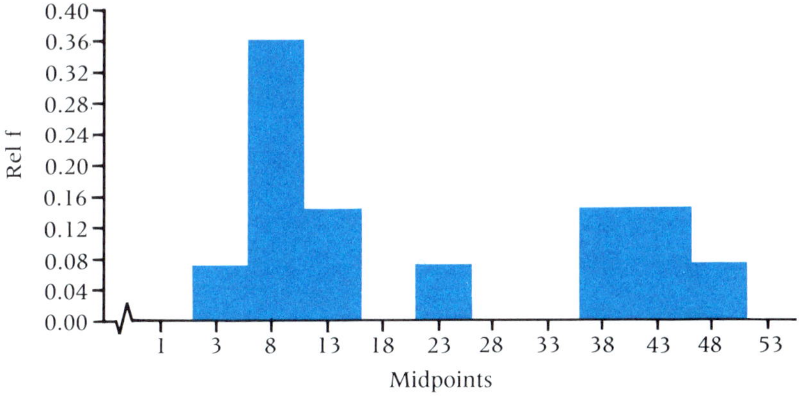
0.40
0.36
0.32
0.28
0.24
0.20
0.16
0.12
0.08
0.04
0.00
Rel f
1 3 8 13 18 23 28 33 38 43 48 53
Midpoints

5.

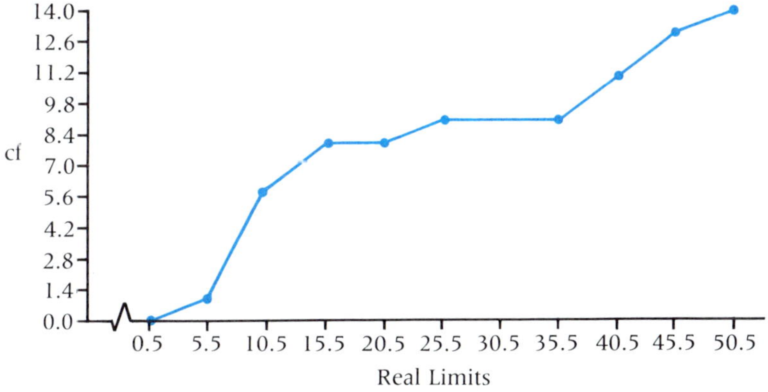
14.0
12.6
11.2
9.8
8.4
7.0
5.6
4.2
2.8
1.4
0.0
cf
0.5 5.5 10.5 15.5 20.5 25.5 30.5 35.5 40.5 45.5 50.5
Real Limits

6. a) 1.95 to 2.05 b) 2.55 to 2.65 c) 1.5 to 4.5 d) 24.5 to 29.5
e) 24.45 to 29.55 f) 24.495 to 29.505

7. Bar graph-rel frequency

8.
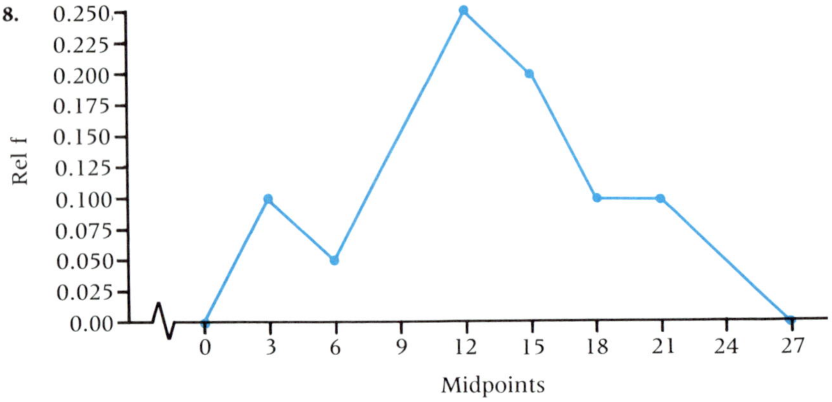

9.
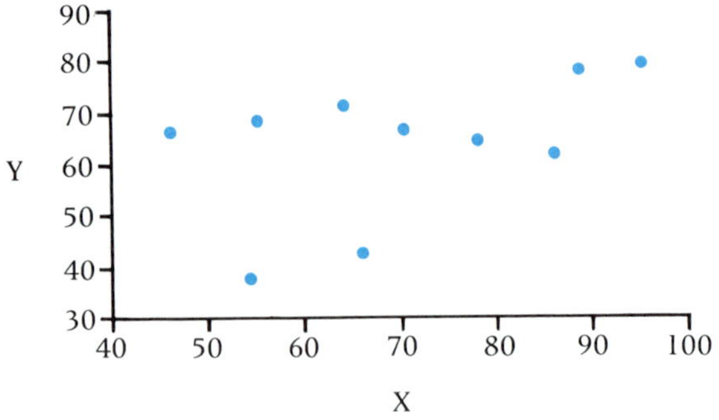

10. Frequency polygon or histogram with relative frequency on the ordinate. On the abscissa should be number of episodes per week in intervals such as 0-2, 3-5, 6-8 etc. with psychotics and neurotics differentiated.

Chapter 3

1.

	Data Set A	Data Set B
μ	7	3
Mo	8	2
Mdn	7.67	2.25

2. $\mu_c = 5.44$

3. Data Set A—negative skew
Data Set B—positive skew

4. Data Set A

$\mu = 6.79$
Mo = 5
Mdn = 6.2
Positive skew

Data Set B

$\mu = 113.85$
Mo = 107
Mdn = 113.14
Positive skew

5. $\mu = 10$ Mo = 11 Mdn = 10.14 Negative skew

6. -5

7. **a)** It is limited at zero. **b)** Mean

Chapter 4

1. Data Set A

X	x
2	−2
4	0
6	2
7	3
4	0
3	−1
2	−2

$\mu = 4$

Data Set B

X	x
12	−16
23	−5
50	22
35	7
20	−8

$\mu = 28$

2. Data Set A

$\mu = 5.86$
$\sigma^2 = 6.98$
$\sigma = 2.64$
Range = 9

Data Set B

$\mu = 3$
$\sigma^2 = 7.25$
$\sigma = 2.69$
Range = 9

3. Data Set A

$\sigma^2 = 7.25$
$\sigma = 2.69$
Range = 9

Data Set B

$\sigma^2 = 3.22$
$\sigma = 1.79$
Range = 6

4. $\sigma^2 = 30.85$
$\sigma = 5.55$

5. $\sigma^2_c = 9.16$

6. distribution A

Chapter 5

1. $P_{50} = 121.5$
$P_{75} = 129.5$
$P_{90} = 133.25$
$PR_{117} = 30.83$

2. $Z = -2$

3.

X	Z
4	−1.02
6	−0.30
8	0.42
9	0.78
11	1.49
3	−1.37

4. Mary did better on her math test
5. A positively skewed distribution
6. 62% of the scores were at or below a score value of 89%.
7. d.
8. X = 63
9. X = 67.9

Chapter 6

1. 0.9876
2. 0.9386
3. 0.0250
4. ±2.58
5. +1.28
6. **a)** 0.50 **b)** 2.5 **c)** 50 **d)** 95 **e)** 99
7. **a)** Area = 25.14% **b)** Area = 31% **c)** 61.92 and 82.08 **d)** 52.35 **e)** 250 **f)** ~109 **g)** 63.96

Chapter 7

1. a,c,d
2. a,c,d
3. **a)** p = 0.56 **b)** p = 0.94
4. **a)** p = 0.04 **b)** p = 0.21 **c)** p = 0.79 **d)** p = 0.11 **e)** no
5. **a)** p = 0.025 **b)** p = 0.325
6. p = 0.32
7. 720
8. 56
9. 5040

Chapter 8

1. $\mu = 5$ $\sigma = 3.63$
2. Samples

1,1	1,2	1,4	1,7	1,11
2,1	2,2	2,4	2,7	2,11
4,1	4,2	4,4	4,7	4,11
7,1	7,2	7,4	7,7	7,11
11,1	11,2	11,4	11,7	11,11

Sample Means

1.00	1.50	2.50	4.00	6.00
1.50	2.00	3.00	4.50	6.50
2.50	3.00	4.00	5.50	7.50
4.00	4.50	5.50	7.00	9.00
6.00	6.50	7.50	9.00	11.00

3.

$\overline{X}$	f	$f\overline{X}$	$\overline{X}^2$	$f\overline{X}^2$
11.00	1	11	121.00	121.0
9.00	2	18	81.00	162.0
7.50	2	15	56.30	112.5
7.00	1	7	49.00	49.0
6.50	2	13	42.30	84.5
6.00	2	12	36.00	72.0
5.50	2	11	30.30	60.5
4.50	2	9	20.30	40.5
4.00	3	12	16.00	48.0
3.00	2	6	9.00	18.0
2.50	2	5	6.25	12.5
2.00	1	2	4.00	4.0
1.50	2	3	2.25	4.5
1.00	1	1	1.00	1.0
	25	125		790

$\mu_{\overline{X}} = 5.0$ $\sigma_{\overline{X}} = 2.57$

4. **a)** $p = 0$ **b)** $p = 12/25$ **c)** $p = 4/25$

5. **a)** $\mu_{\overline{X}} = 75$ $\sigma_{\overline{X}} = 1$
b) $\mu_{\overline{X}} = 75$ $\sigma_{\overline{X}} = 1.5$
c) $\mu_{\overline{X}} = 75$ $\sigma_{\overline{X}} = 3$

6. $p = 0.9050$

Chapter 9

1. H_o: $\mu = 79$ H_a: $\mu \neq 79$
$Z = 2.14$ Fail to reject

2. 95% CI: 79.51 to 90.49
99% CI: 77.78 to 92.22

3. H_o: $P = 0.50$ H_a: $P > 0.50$
$Z = 0.86$ Fail to reject

4. H_o: $\mu = 250$ H_a: $\mu \neq 250$
$Z = -10.00$ Reject

5. H_o: $P = 0.75$ H_a: $P > 0.75$
$Z = 1.39$ Fail to reject

6. 99%CI: 42.84 to 53.16

7. H_o: $P = 0.70$ H_a: $P \neq 0.70$
$Z = -1.54$ Fail to reject

8. H_o: $P = 0.50$ H_a: $P > 0.50$
$Z = 1.76$ Fail to reject

Chapter 10

1. t-test for Dependent Means

2. t-test for Dependent Means

3. t-test for Independent Means

4. **a)** 28 **b)** 15

5. $H_o: \mu_A = \mu_B$
$H_a: \mu_A \neq \mu_B$
$t_{.01} = \pm 2.977$
$t = -1.03$ Fail to reject

6. $H_o: \mu_1 = \mu_2$
$H_a: \mu_1 \neq \mu_2$
$t_{.01} = \pm 2.98$
$t = -2.323$ Fail to reject

7. $H_o: \mu_1 = \mu_2$
$H_a: \mu_1 \neq \mu_2$
$t_{.05} = \pm 2.145$
$t = 3.813$ Reject

8. $C(1.38 \leq \mu_1 - \mu_2 \leq 8.62) = .95$

9. **a)** Rejecting a true null hypothesis. **b)** Failing to reject a false null hypothesis. **c)** The probability of rejecting a false null hypothesis.

10. **1)** Use a dependent samples design to decrease the error variance. **2)** Use a larger alpha level. **3)** Increase sample size.

Chapter 11

1. F = 3.54
Reject

2. F = 24.27
Reject

3. F = 14.39
Reject

4.

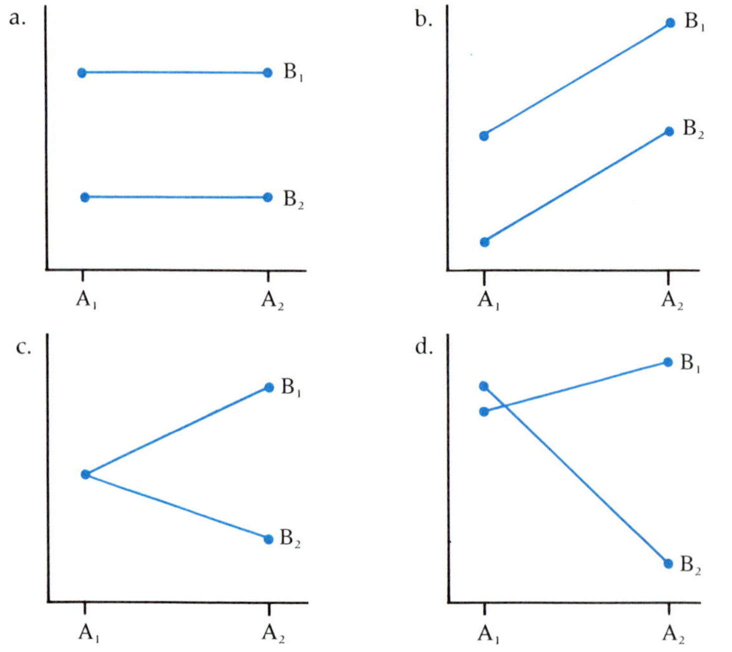

5.

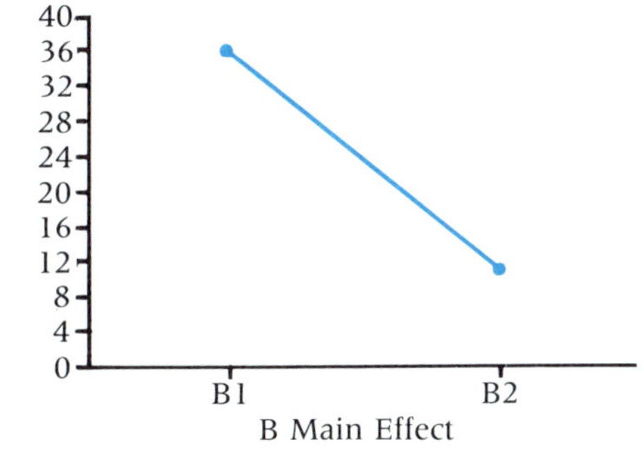

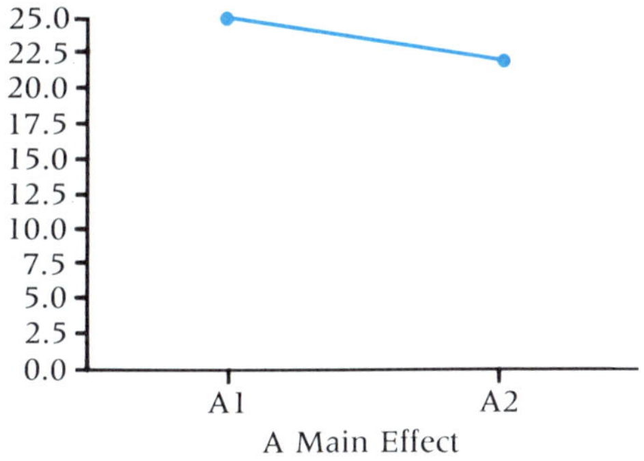

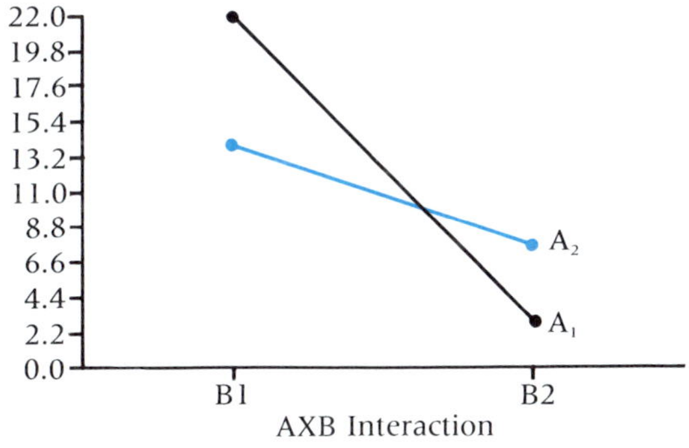

6.

	F's	p
A	21.14	<.01
B	21.14	<.01
AXB	9.40	<.01

Chapter 12

1.

Source	df
a) Total	99
b) Subject	19
c) Within subject	80
d) Between treatment	4
e) Subject by treatment	76

2. F = 7.16, $p < .01$

3.

	F's	p
A	2.67	<.05
B	11.20	<.01
AXB	2.93	>.05

4.

	F's	p
A	0.08	>.05
B	18.18	<.01
AXB	19.88	<.01

5. Graphs of main effects and interaction from 4.

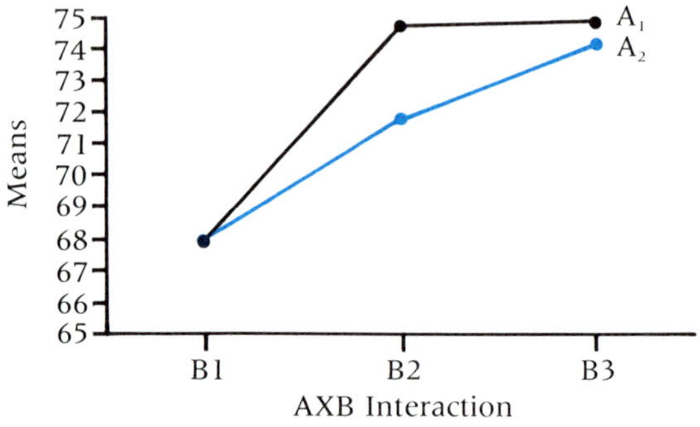

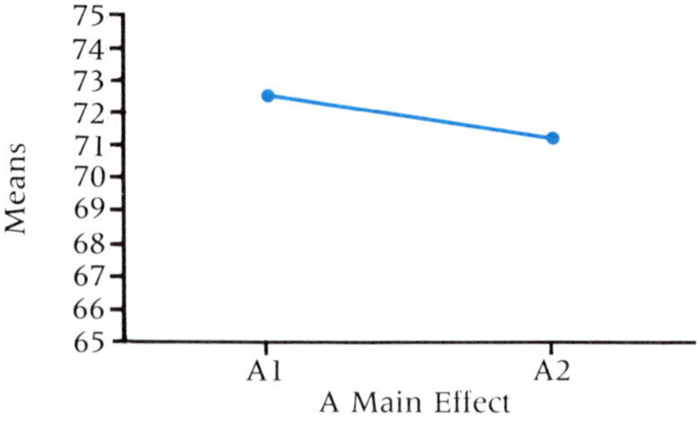

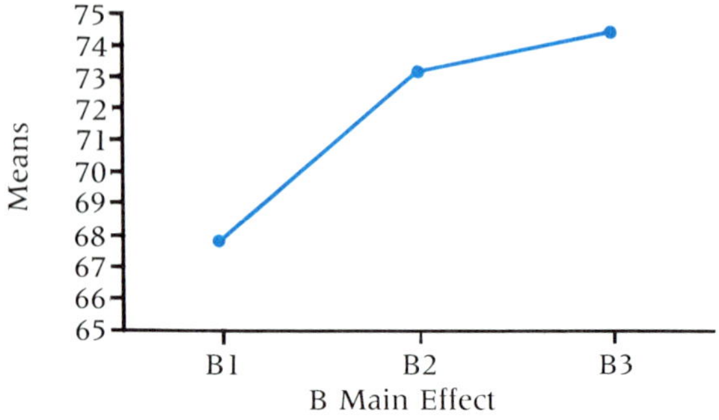

Chapter 13

1. $F_s = 2.79$

a) $C = 1.6$ $s_c = 1.33$ $F' = 1.20$

Fail to reject

b) $C = 4.20$ $s_c = 1.15$

$F' = 3.65$ Reject

c) $C = 5.00$ $s_c = 1.33$ $F' = 3.76$ Reject

2. $q = 3.95$

ANOVA

$SS_{TOT} = 78.25$

$SS_{BG} = 50$

$SS_{WG} = 28.25$

$MS_{WG} = 3.14$

$HSD = 3.50$

$\overline{X}_c - \overline{X}_a = 5.00$*

$\overline{X}_c - \overline{X}_b = 2.50$

$\overline{X}_b - \overline{X}_a = 2.5$

*Significantly different

3.

	SS	df	MS	F
Between	28.40	2	14.20	9.26
Within	9.20	6	1.53	

	1 vs 2	1 vs 3	3 vs 2	2 & 3 vs 1	1 & 3 vs 2	1 & 2 vs 3
C	4.6	3.9	0.7	4.25	2.65	1.60
s_c	1.01	1.01	1.01	0.87	0.87	0.87
F′	4.55*	3.86*	0.69	4.89*	3.05	1.84

$F_s = 3.21$ *significant at $\alpha = .05$

4. a)

	SS	df	MS	F	p
Between	9.66	3	3.22	28.78	<.01
Within	2.24	20	0.11		
Total	11.9	23			

$F_s = 3.05$

b) Mirage vs Eclipse

$C = 0.79$

$s_c = 0.19$

$F' = 4.16$*

c) Average of Mirage, Dancer and Mark IV vs Eclipse

$C = 1.19$

$s_c = 0.10$

$F' = 11.9$*

*Significantly different

d) Tukey's HSD = 0.54

Mirage vs Eclipse = 0.78*

Dancer vs Eclipse = 0.98*

Mark IV vs Eclipse = 1.78*

Dancer vs Mirage = 0.20

Mark IV vs Mirage = 0.20

Mark IV vs Mirage = 1.00*

Mark IV vs Dancer = 0.80*

*Significantly different

Chapter 14

1. a) df = 5 $\chi^2_{.05} = 11.07$
   b) df = 1 $\chi^2_{.01} = 6.64$
   c) df = 12 $\chi^2_{.05} = 21.03$
   d) df = 1 $\chi^2_{.01} = 6.64$
2. $\chi^2 = 4.3$, Fail to reject
3. $\chi^2 = 42.67$, Reject
4. $\chi^2 = 8$, Reject
5. $\chi^2 = 0.04$, Fail to reject
6. $\chi^2 = 18.3$, Reject
7. $\chi^2 = 28.4$, Reject
8. $\chi^2 = 4.19$, Fail to reject
9. $\chi^2 = 12.8$, Reject
10. $\chi^2 = 35.88$, Reject

Chapter 15

1. $U_1 = 4$ $U_2 = 60$
   $U_{CRIT} = 7$, Reject
2. T = 14.5
   $T_{CRIT} = 3$
   Do not reject
3. H = 8.42
   Critical Value of Chi Square = 5.99
   Reject
4. rho = 0.86
   $rho_{crit} = 0.65$
   Reject

Chapter 16

1.

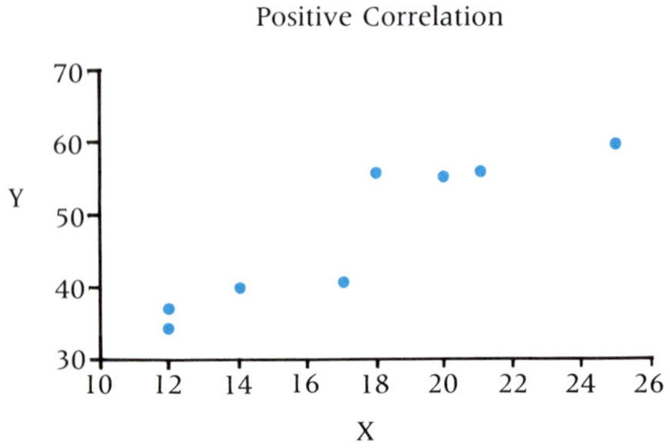

2.

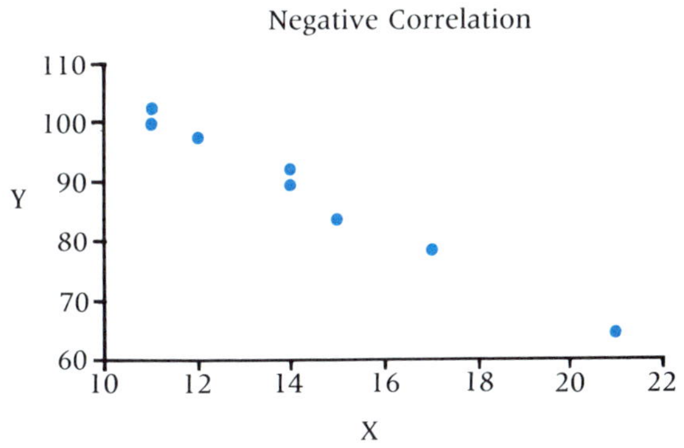

3. $\rho = 0.93$

4. $\rho = -0.99$
$\rho^2 = 0.98$

5. $\rho = 0.88$

6. $H_0: \rho = 0$
$H_a: \rho \neq 0$
$t_{.05} = \pm 2.16$
$s_r = 0.21$
$t = 3.00$

7. $\rho = -0.81$

Chapter 17

1. $b = 3.78$
$\mu_x = 5.53$
$\mu_y = 60$

2. $a = 39.11$

3. & 4. Regression Line

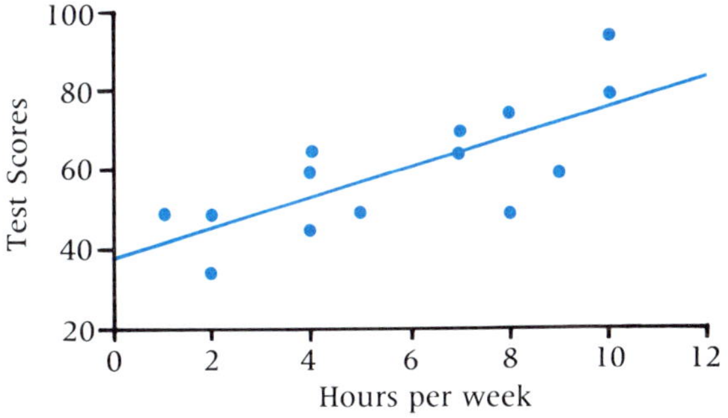

5. $Y' = 67.43$

6. **a)** $Y' = 6.46$ **b)** $Y' = 11.50$ **c)** $Y' = 8.98$

7. **a)** 0.3 **b)** −0.9 **c)** 0

Appendix B
Statistical Tables

Table 1 Areas Under the Normal Curve

z	 0 z	 0 z
0.00	.0000	.5000
0.01	.0040	.4960
0.02	.0080	.4920
0.03	.0120	.4880
0.04	.0160	.4840
0.05	.0199	.4801
0.06	.0239	.4761
0.07	.0279	.4721
0.08	.0319	.4681
0.09	.0359	.4641
0.10	.0398	.4602
0.11	.0438	.4562
0.12	.0478	.4522
0.13	.0517	.4483
0.14	.0557	.4443
0.15	.0596	.4404
0.16	.0636	.4364
0.17	.0675	.4325
0.18	.0714	.4286
0.19	.0753	.4247
0.20	.0793	.4207
0.21	.0832	.4168
0.22	.0871	.4129
0.23	.0910	.4090
0.24	.0948	.4052
0.25	.0987	.4013
0.26	.1026	.3974
0.27	.1064	.3936
0.28	.1103	.3897
0.29	.1141	.3859
0.30	.1179	.3821
0.31	.1217	.3783
0.32	.1255	.3745
0.33	.1293	.3707
0.34	.1331	.3669
0.35	.1368	.3632
0.36	.1406	.3594
0.37	.1443	.3557
0.38	.1480	.3520
0.39	.1517	.3483
0.40	.1554	.3446
0.41	.1591	.3409
0.42	.1628	.3372
0.43	.1664	.3336
0.44	.1700	.3300
0.45	.1736	.3264
0.46	.1772	.3228
0.47	.1808	.3192
0.48	.1844	.3156
0.49	.1879	.3121
0.50	.1915	.3085
0.51	.1950	.3050
0.52	.1985	.3015
0.53	.2019	.2981
0.54	.2054	.2946

z	 0 z	 0 z
0.55	.2088	.2912
0.56	.2123	.2877
0.57	.2157	.2843
0.58	.2190	.2810
0.59	.2224	.2776
0.60	.2257	.2743
0.61	.2291	.2709
0.62	.2324	.2676
0.63	.2357	.2643
0.64	.2389	.2611
0.65	.2422	.2578
0.66	.2454	.2546
0.67	.2486	.2514
0.68	.2517	.2483
0.69	.2549	.2451
0.70	.2580	.2420
0.71	.2611	.2389
0.72	.2642	.2358
0.73	.2673	.2327
0.74	.2704	.2296
0.75	.2734	.2266
0.76	.2764	.2236
0.77	.2794	.2206
0.78	.2823	.2177
0.79	.2852	.2148
0.80	.2881	.2119
0.81	.2910	.2090
0.82	.2939	.2061
0.83	.2967	.2033
0.84	.2995	.2005
0.85	.3023	.1977
0.86	.3051	.1949
0.87	.3078	.1922
0.88	.3106	.1894
0.89	.3133	.1867
0.90	.3159	.1841
0.91	.3186	.1814
0.92	.3212	.1788
0.93	.3238	.1762
0.94	.3264	.1736
0.95	.3289	.1711
0.96	.3315	.1685
0.97	.3340	.1660
0.98	.3365	.1635
0.99	.3389	.1611
1.00	.3413	.1587
1.01	.3438	.1562
1.02	.3461	.1539
1.03	.3485	.1515
1.04	.3508	.1492
1.05	.3531	.1469
1.06	.3554	.1446
1.07	.3577	.1423
1.08	.3599	.1401
1.09	.3621	.1379

z	 0 z	 0 z
1.10	.3643	.1357
1.11	.3665	.1335
1.12	.3686	.1314
1.13	.3708	.1292
1.14	.3729	.1271
1.15	.3749	.1251
1.16	.3770	.1230
1.17	.3790	.1210
1.18	.3810	.1190
1.19	.3830	.1170
1.20	.3849	.1151
1.21	.3869	.1131
1.22	.3888	.1112
1.23	.3907	.1093
1.24	.3925	.1075
1.25	.3944	.1056
1.26	.3962	.1038
1.27	.3980	.1020
1.28	.3997	.1003
1.29	.4015	.0985
1.30	.4032	.0968
1.31	.4049	.0951
1.32	.4066	.0934
1.33	.4082	.0918
1.34	.4099	.0901
1.35	.4115	.0885
1.36	.4131	.0869
1.37	.4147	.0853
1.38	.4162	.0838
1.39	.4177	.0823
1.40	.4192	.0808
1.41	.4207	.0793
1.42	.4222	.0778
1.43	.4236	.0764
1.44	.4251	.0749
1.45	.4265	.0735
1.46	.4279	.0721
1.47	.4292	.0708
1.48	.4306	.0694
1.49	.4319	.0681
1.50	.4332	.0668
1.51	.4345	.0655
1.52	.4357	.0643
1.53	.4370	.0630
1.54	.4382	.0618
1.55	.4394	.0606
1.56	.4406	.0594
1.57	.4418	.0582
1.58	.4429	.0571
1.59	.4441	.0559
1.60	.4452	.0548
1.61	.4463	.0537
1.62	.4474	.0526
1.63	.4484	.0516
1.64	.4495	.0505

Table 1 (cont'd)

z	 0 z	 0 z	z	 0 z	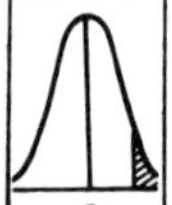 0 z	z	 0 z	 0 z
1.65	.4505	.0495	2.22	.4868	.0132	2.79	.4974	.0026
1.66	.4515	.0485	2.23	.4871	.0129	2.80	.4974	.0026
1.67	.4525	.0475	2.24	.4875	.0125	2.81	.4975	.0025
1.68	.4535	.0465	2.25	.4878	.0122	2.82	.4976	.0024
1.69	.4545	.0455	2.26	.4881	.0119	2.83	.4977	.0023
1.70	.4554	.0446	2.27	.4884	.0116	2.84	.4977	.0023
1.71	.4564	.0436	2.28	.4887	.0113	2.85	.4978	.0022
1.72	.4573	.0427	2.29	.4890	.0110	2.86	.4979	.0021
1.73	.4582	.0418	2.30	.4893	.0107	2.87	.4979	.0021
1.74	.4591	.0409	2.31	.4896	.0104	2.88	.4980	.0020
1.75	.4599	.0401	2.32	.4898	.0102	2.89	.4981	.0019
1.76	.4608	.0392	2.33	.4901	.0099	2.90	.4981	.0019
1.77	.4616	.0384	2.34	.4904	.0096	2.91	.4982	.0018
1.78	.4625	.0375	2.35	.4906	.0094	2.92	.4982	.0018
1.79	.4633	.0367	2.36	.4909	.0091	2.93	.4983	.0017
1.80	.4641	.0359	2.37	.4911	.0089	2.94	.4984	.0016
1.81	.4649	.0351	2.38	.4913	.0087	2.95	.4984	.0016
1.82	.4656	.0344	2.39	.4916	.0084	2.96	.4985	.0015
1.83	.4664	.0336	2.40	.4918	.0082	2.97	.4985	.0015
1.84	.4671	.0329	2.41	.4920	.0080	2.98	.4986	.0014
1.85	.4678	.0322	2.42	.4922	.0078	2.99	.4986	.0014
1.86	.4686	.0314	2.43	.4925	.0075	3.00	.4987	.0013
1.87	.4693	.0307	2.44	.4927	.0073	3.01	.4987	.0013
1.88	.4699	.0301	2.45	.4929	.0071	3.02	.4987	.0013
1.89	.4706	.0294	2.46	.4931	.0069	3.03	.4988	.0012
1.90	.4713	.0287	2.47	.4932	.0068	3.04	.4988	.0012
1.91	.4719	.0281	2.48	.4934	.0066	3.05	.4989	.0011
1.92	.4726	.0274	2.49	.4936	.0064	3.06	.4989	.0011
1.93	.4732	.0268	2.50	.4938	.0062	3.07	.4989	.0011
1.94	.4738	.0262	2.51	.4940	.0060	3.08	.4990	.0010
1.95	.4744	.0256	2.52	.4941	.0059	3.09	.4990	.0010
1.96	.4750	.0250	2.53	.4943	.0057	3.10	.4990	.0010
1.97	.4756	.0244	2.54	.4945	.0055	3.11	.4991	.0009
1.98	.4761	.0239	2.55	.4946	.0054	3.12	.4991	.0009
1.99	.4767	.0233	2.56	.4948	.0052	3.13	.4991	.0009
2.00	.4772	.0228	2.57	.4949	.0051	3.14	.4992	.0008
2.01	.4778	.0222	2.58	.4951	.0049	3.15	.4992	.0008
2.02	.4783	.0217	2.59	.4952	.0048	3.16	.4992	.0008
2.03	.4788	.0212	2.60	.4953	.0047	3.17	.4992	.0008
2.04	.4793	.0207	2.61	.4955	.0045	3.18	.4993	.0007
2.05	.4798	.0202	2.62	.4956	.0044	3.19	.4993	.0007
2.06	.4803	.0197	2.63	.4957	.0043	3.20	.4993	.0007
2.07	.4808	.0192	2.64	.4959	.0041	3.21	.4993	.0007
2.08	.4812	.0188	2.65	.4960	.0040	3.22	.4994	.0006
2.09	.4817	.0183	2.66	.4961	.0039	3.23	.4994	.0006
2.10	.4821	.0179	2.67	.4962	.0038	3.24	.4994	.0006
2.11	.4826	.0174	2.68	.4963	.0037	3.25	.4994	.0006
2.12	.4830	.0170	2.69	.4964	.0036	3.30	.4995	.0005
2.13	.4834	.0166	2.70	.4965	.0035	3.35	.4996	.0004
2.14	.4838	.0162	2.71	.4966	.0034	3.40	.4997	.0003
2.15	.4842	.0158	2.72	.4967	.0033	3.45	.4997	.0003
2.16	.4846	.0154	2.73	.4968	.0032	3.50	.4998	.0002
2.17	.4850	.0150	2.74	.4969	.0031	3.60	.4998	.0002
2.18	.4854	.0146	2.75	.4970	.0030	3.70	.4999	.0001
2.19	.4857	.0143	2.76	.4971	.0029	3.80	.4999	.0001
2.20	.4861	.0139	2.77	.4972	.0028	3.90	.49995	.00005
2.21	.4864	.0136	2.78	.4973	.0027	4.00	.49997	.00003

Source: Richard Runyon and Audrey Haber, *Fundamentals of Behavioral Statistics*, 2ed. (New York, N.Y.: Random House, Inc. 1971).

Table 2 Critical Values of t

	Level of significance for a directional (one-tailed) test					
	.10	.05	.025	.01	.005	.0005
	Level of significance for a non-directional (two-tailed) test					
df	.20	.10	.05	.02	.01	.001
1	3.078	6.314	12.706	31.821	63.657	636.619
2	1.886	2.920	4.303	6.965	9.925	31.598
3	1.638	2.353	3.182	4.541	5.841	12.941
4	1.533	2.132	2.776	3.747	4.604	8.610
5	1.476	2.015	2.571	3.365	4.032	6.859
6	1.440	1.943	2.447	3.143	3.707	5.959
7	1.415	1.895	2.365	2.998	3.499	5.405
8	1.397	1.860	2.306	2.896	3.355	5.041
9	1.383	1.833	2.262	2.821	3.250	4.781
10	1.372	1.812	2.228	2.764	3.169	4.587
11	1.363	1.796	2.201	2.718	3.106	4.437
12	1.356	1.782	2.179	2.681	3.055	4.318
13	1.350	1.771	2.160	2.650	3.012	4.221
14	1.345	1.761	2.145	2.624	2.977	4.140
15	1.341	1.753	2.131	2.602	2.947	4.073
16	1.337	1.746	2.120	2.583	2.921	4.015
17	1.333	1.740	2.110	2.567	2.898	3.965
18	1.330	1.734	2.101	2.552	2.878	3.922
19	1.328	1.729	2.093	2.539	2.861	3.883
20	1.325	1.725	2.086	2.528	2.845	3.850
21	1.323	1.721	2.080	2.518	2.831	3.819
22	1.321	1.717	2.074	2.508	2.819	3.792
23	1.319	1.714	2.069	2.500	2.807	3.767
24	1.318	1.711	2.064	2.492	2.797	3.745
25	1.316	1.708	2.060	2.485	2.787	3.725
26	1.315	1.706	2.056	2.479	2.779	3.707
27	1.314	1.703	2.052	2.473	2.771	3.690
28	1.313	1.701	2.048	2.467	2.763	3.674
29	1.311	1.699	2.045	2.462	2.756	3.659
30	1.310	1.697	2.042	2.457	2.750	3.646
40	1.303	1.684	2.021	2.423	2.704	3.551
60	1.296	1.671	2.000	2.390	2.660	3.460
120	1.289	1.658	1.980	2.358	2.617	3.373
∞	1.282	1.645	1.960	2.326	2.576	3.291

Source: From *Fundamental Statistics for Psychology,* copyright © 1975 by Robert McCall, reproduced by permission of Harcourt Brace Jovanovich, Inc.

Table 3 Critical Values of F

(.05 level in roman type, **.01 level in bold face**)

Degrees of freedom for the denominator	Degrees of freedom for the numerator 1	2	3	4	5	6	7	8	9	10	11	12	14	16	20	24	30	40	50	75	100	200	500	∞	
1	161 **4,052**	200 **4,999**	216 **5,403**	225 **5,625**	230 **5,764**	234 **5,859**	237 **5,928**	239 **5,981**	241 **6,022**	242 **6,056**	243 **6,082**	244 **6,106**	245 **6,142**	246 **6,169**	248 **6,208**	249 **6,234**	250 **6,261**	251 **6,286**	252 **6,302**	253 **6,323**	253 **6,334**	254 **6,352**	254 **6,361**	254 **6,366**	1
2	18.51 **98.49**	19.00 **99.00**	19.16 **99.17**	19.25 **99.25**	19.30 **99.30**	19.33 **99.33**	19.36 **99.36**	19.37 **99.37**	19.38 **99.39**	19.39 **99.40**	19.40 **99.41**	19.41 **99.42**	19.42 **99.43**	19.43 **99.44**	19.44 **99.45**	19.45 **99.46**	19.46 **99.47**	19.47 **99.48**	19.47 **99.48**	19.48 **99.49**	19.49 **99.49**	19.49 **99.49**	19.50 **99.50**	19.50 **99.50**	2
3	10.13 **34.12**	9.55 **30.82**	9.28 **29.46**	9.12 **28.71**	9.01 **28.24**	8.94 **27.91**	8.88 **27.67**	8.84 **27.49**	8.81 **27.34**	8.78 **27.23**	8.76 **27.13**	8.74 **27.05**	8.71 **26.92**	8.69 **26.83**	8.66 **26.69**	8.64 **26.60**	8.62 **26.50**	8.60 **26.41**	8.58 **26.35**	8.57 **26.27**	8.56 **26.23**	8.54 **26.18**	8.54 **26.14**	8.53 **26.12**	3
4	7.71 **21.20**	6.94 **18.00**	6.59 **16.69**	6.39 **15.98**	6.26 **15.52**	6.16 **15.21**	6.09 **14.98**	6.04 **14.80**	6.00 **14.66**	5.96 **14.54**	5.93 **14.45**	5.91 **14.37**	5.87 **14.24**	5.84 **14.15**	5.80 **14.02**	5.77 **13.93**	5.74 **13.83**	5.71 **13.74**	5.70 **13.69**	5.68 **13.61**	5.66 **13.57**	5.65 **13.52**	5.64 **13.48**	5.63 **13.46**	4
5	6.61 **16.26**	5.79 **13.27**	5.41 **12.06**	5.19 **11.39**	5.05 **10.97**	4.95 **10.67**	4.88 **10.45**	4.82 **10.29**	4.78 **10.15**	4.74 **10.05**	4.70 **9.96**	4.68 **9.89**	4.64 **9.77**	4.60 **9.68**	4.56 **9.55**	4.53 **9.47**	4.50 **9.38**	4.46 **9.29**	4.44 **9.24**	4.42 **9.17**	4.40 **9.13**	4.38 **9.07**	4.37 **9.04**	4.36 **9.02**	5
6	5.99 **13.74**	5.14 **10.92**	4.76 **9.78**	4.53 **9.15**	4.39 **8.75**	4.28 **8.47**	4.21 **8.26**	4.15 **8.10**	4.10 **7.98**	4.06 **7.87**	4.03 **7.79**	4.00 **7.72**	3.96 **7.60**	3.92 **7.52**	3.87 **7.39**	3.84 **7.31**	3.81 **7.23**	3.77 **7.14**	3.75 **7.09**	3.72 **7.02**	3.71 **6.99**	3.69 **6.94**	3.68 **6.90**	3.67 **6.88**	6
7	5.59 **12.25**	4.74 **9.55**	4.35 **8.45**	4.12 **7.85**	3.97 **7.46**	3.87 **7.19**	3.79 **7.00**	3.73 **6.84**	3.68 **6.71**	3.63 **6.62**	3.60 **6.54**	3.57 **6.47**	3.52 **6.35**	3.49 **6.27**	3.44 **6.15**	3.41 **6.07**	3.38 **5.98**	3.34 **5.90**	3.32 **5.85**	3.29 **5.78**	3.28 **5.75**	3.25 **5.70**	3.24 **5.67**	3.23 **5.65**	7
8	5.32 **11.26**	4.46 **8.65**	4.07 **7.59**	3.84 **7.01**	3.69 **6.63**	3.58 **6.37**	3.50 **6.19**	3.44 **6.03**	3.39 **5.91**	3.34 **5.82**	3.31 **5.74**	3.28 **5.67**	3.23 **5.56**	3.20 **5.48**	3.15 **5.36**	3.12 **5.28**	3.08 **5.20**	3.05 **5.11**	3.03 **5.06**	3.00 **5.00**	2.98 **4.96**	2.96 **4.91**	2.94 **4.88**	2.93 **4.86**	8
9	5.12 **10.56**	4.26 **8.02**	3.86 **6.99**	3.63 **6.42**	3.48 **6.06**	3.37 **5.80**	3.29 **5.62**	3.23 **5.47**	3.18 **5.35**	3.13 **5.26**	3.10 **5.18**	3.07 **5.11**	3.02 **5.00**	2.98 **4.92**	2.93 **4.80**	2.90 **4.73**	2.86 **4.64**	2.82 **4.56**	2.80 **4.51**	2.77 **4.45**	2.76 **4.41**	2.73 **4.36**	2.72 **4.33**	2.71 **4.31**	9
10	4.96 **10.04**	4.10 **7.56**	3.71 **6.55**	3.48 **5.99**	3.33 **5.64**	3.22 **5.39**	3.14 **5.21**	3.07 **5.06**	3.02 **4.95**	2.97 **4.85**	2.94 **4.78**	2.91 **4.71**	2.86 **4.60**	2.82 **4.52**	2.77 **4.41**	2.74 **4.33**	2.70 **4.25**	2.67 **4.17**	2.64 **4.12**	2.61 **4.05**	2.59 **4.01**	2.56 **3.96**	2.55 **3.93**	2.54 **3.91**	10
11	4.84 **9.65**	3.98 **7.20**	3.59 **6.22**	3.36 **5.67**	3.20 **5.32**	3.09 **5.07**	3.01 **4.88**	2.95 **4.74**	2.90 **4.63**	2.86 **4.54**	2.82 **4.46**	2.79 **4.40**	2.74 **4.29**	2.70 **4.21**	2.65 **4.10**	2.61 **4.02**	2.57 **3.94**	2.53 **3.86**	2.50 **3.80**	2.47 **3.74**	2.45 **3.70**	2.42 **3.66**	2.41 **3.62**	2.40 **3.60**	11
12	4.75 **9.33**	3.88 **6.93**	3.49 **5.95**	3.26 **5.41**	3.11 **5.06**	3.00 **4.82**	2.92 **4.65**	2.85 **4.50**	2.80 **4.39**	2.76 **4.30**	2.72 **4.22**	2.69 **4.16**	2.64 **4.05**	2.60 **3.98**	2.54 **3.86**	2.50 **3.78**	2.46 **3.70**	2.42 **3.61**	2.40 **3.56**	2.36 **3.49**	2.35 **3.46**	2.32 **3.41**	2.31 **3.38**	2.30 **3.36**	12
13	4.67 **9.07**	3.80 **6.70**	3.41 **5.74**	3.18 **5.20**	3.02 **4.86**	2.92 **4.62**	2.84 **4.44**	2.77 **4.30**	2.72 **4.19**	2.67 **4.10**	2.63 **4.02**	2.60 **3.96**	2.55 **3.85**	2.51 **3.78**	2.46 **3.67**	2.42 **3.59**	2.38 **3.51**	2.34 **3.42**	2.32 **3.37**	2.28 **3.30**	2.26 **3.27**	2.24 **3.21**	2.22 **3.18**	2.21 **3.16**	13

Table 3 (cont'd)

Degrees of freedom for the denominator	Degrees of freedom for the numerator																								
	1	2	3	4	5	6	7	8	9	10	11	12	14	16	20	24	30	40	50	75	100	200	500	∞	
14	4.60 **8.86**	3.74 **6.51**	3.34 **5.56**	3.11 **5.03**	2.96 **4.69**	2.85 **4.46**	2.77 **4.28**	2.70 **4.14**	2.65 **4.03**	2.60 **3.94**	2.56 **3.86**	2.53 **3.80**	2.48 **3.70**	2.44 **3.62**	2.39 **3.51**	2.35 **3.43**	2.31 **3.34**	2.27 **3.26**	2.24 **3.21**	2.21 **3.14**	2.19 **3.11**	2.16 **3.06**	2.14 **3.02**	2.13 **3.00**	14
15	4.54 **8.68**	3.68 **6.36**	3.29 **5.42**	3.06 **4.89**	2.90 **4.56**	2.79 **4.32**	2.70 **4.14**	2.64 **4.00**	2.59 **3.89**	2.55 **3.80**	2.51 **3.73**	2.48 **3.67**	2.43 **3.56**	2.39 **3.48**	2.33 **3.36**	2.29 **3.29**	2.25 **3.20**	2.21 **3.12**	2.18 **3.07**	2.15 **3.00**	2.12 **2.97**	2.10 **2.92**	2.08 **2.89**	2.07 **2.87**	15
16	4.49 **8.53**	3.63 **6.23**	3.24 **5.29**	3.01 **4.77**	2.85 **4.44**	2.74 **4.20**	2.66 **4.03**	2.59 **3.89**	2.54 **3.78**	2.49 **3.69**	2.45 **3.61**	2.42 **3.55**	2.37 **3.45**	2.33 **3.37**	2.28 **3.25**	2.24 **3.18**	2.20 **3.10**	2.16 **3.01**	2.13 **2.96**	2.09 **2.98**	2.07 **2.86**	2.04 **2.80**	2.02 **2.77**	2.01 **2.75**	16
17	4.45 **8.40**	3.59 **6.11**	3.20 **5.18**	2.96 **4.67**	2.81 **4.34**	2.70 **4.10**	2.62 **3.93**	2.55 **3.79**	2.50 **3.68**	2.45 **3.59**	2.41 **3.52**	2.38 **3.45**	2.33 **3.35**	2.29 **3.27**	2.23 **3.16**	2.19 **3.08**	2.15 **3.00**	2.11 **2.92**	2.08 **2.86**	2.04 **2.79**	2.02 **2.76**	1.99 **2.70**	1.97 **2.67**	1.96 **2.65**	17
18	4.41 **8.28**	3.55 **6.01**	3.16 **5.09**	2.93 **4.58**	2.77 **4.25**	3.66 **4.01**	2.58 **3.85**	2.51 **3.71**	2.46 **3.60**	2.41 **3.51**	2.37 **3.44**	2.34 **3.37**	2.29 **3.27**	2.25 **3.19**	2.19 **3.07**	2.15 **3.00**	2.11 **2.91**	2.07 **2.83**	2.04 **2.78**	2.00 **2.71**	1.98 **2.68**	1.95 **2.62**	1.93 **2.59**	1.92 **2.57**	18
19	4.38 **8.18**	3.52 **5.93**	3.13 **5.01**	2.90 **4.50**	2.74 **4.17**	2.63 **3.94**	2.55 **3.77**	2.48 **3.63**	2.43 **3.52**	2.38 **3.43**	2.34 **3.36**	2.31 **3.30**	2.26 **3.19**	2.21 **3.12**	2.15 **3.00**	2.11 **2.92**	2.07 **2.84**	2.02 **2.76**	2.00 **2.70**	1.96 **2.63**	1.94 **2.60**	1.91 **2.54**	1.90 **2.51**	1.88 **2.49**	19
20	4.35 **8.10**	3.49 **5.85**	3.10 **4.94**	2.87 **4.43**	2.71 **4.10**	2.60 **3.87**	2.52 **3.71**	2.45 **3.56**	2.40 **3.45**	2.35 **3.37**	2.31 **3.30**	2.28 **3.23**	2.23 **3.13**	2.18 **3.05**	2.12 **2.94**	2.08 **2.86**	2.04 **2.77**	1.99 **2.69**	1.96 **2.63**	1.92 **2.56**	1.90 **2.53**	1.87 **2.47**	1.85 **2.44**	1.84 **2.42**	20
21	4.32 **8.02**	3.47 **5.78**	3.07 **4.87**	2.84 **4.37**	2.68 **4.04**	2.57 **3.81**	2.49 **3.65**	2.42 **3.51**	2.37 **3.40**	2.32 **3.31**	2.28 **3.24**	2.25 **3.17**	2.20 **3.07**	2.15 **2.99**	2.09 **2.88**	2.05 **2.80**	2.00 **2.72**	1.96 **2.63**	1.93 **2.58**	1.89 **2.51**	1.87 **2.47**	1.84 **2.42**	1.82 **2.38**	1.81 **2.36**	21
22	4.30 **7.94**	3.44 **5.72**	3.05 **4.82**	2.82 **4.31**	2.66 **3.99**	2.55 **3.76**	2.47 **3.59**	2.40 **3.45**	2.35 **3.35**	2.30 **3.26**	2.26 **3.18**	2.23 **3.12**	2.18 **3.02**	2.13 **2.94**	2.07 **2.83**	2.03 **2.75**	1.98 **2.67**	1.93 **2.58**	1.91 **2.53**	1.87 **2.46**	1.84 **2.42**	1.81 **2.37**	1.80 **2.33**	1.78 **2.31**	22
23	4.28 **7.88**	3.42 **5.66**	3.03 **4.76**	2.80 **4.26**	2.64 **3.94**	2.53 **3.71**	2.45 **3.54**	2.38 **3.41**	2.32 **3.30**	2.28 **3.21**	2.24 **3.14**	2.20 **3.07**	2.14 **2.97**	2.10 **2.89**	2.04 **2.78**	2.00 **2.70**	1.96 **2.62**	1.91 **2.53**	1.88 **2.48**	1.84 **2.41**	1.82 **2.37**	1.79 **2.32**	1.77 **2.28**	1.76 **2.26**	23
24	4.26 **7.82**	3.40 **5.61**	3.01 **4.72**	2.78 **4.22**	2.62 **3.90**	2.51 **3.67**	2.43 **3.50**	2.36 **3.36**	2.30 **3.25**	2.26 **3.17**	2.22 **3.09**	2.18 **3.03**	2.13 **2.93**	2.09 **2.85**	2.02 **2.74**	1.98 **2.66**	1.94 **2.58**	1.89 **2.49**	1.86 **2.44**	1.82 **2.36**	1.80 **2.33**	1.76 **2.27**	1.74 **2.23**	1.73 **2.21**	24
25	4.24 **7.77**	3.38 **5.57**	2.99 **4.68**	2.76 **4.18**	2.60 **3.86**	2.49 **3.63**	2.41 **3.46**	2.34 **3.32**	2.28 **3.21**	2.24 **3.13**	2.20 **3.05**	2.16 **2.99**	2.11 **2.89**	2.06 **2.81**	2.00 **2.70**	1.96 **2.62**	1.92 **2.54**	1.87 **2.45**	1.84 **2.40**	1.80 **2.32**	1.77 **2.29**	1.74 **2.23**	1.72 **2.19**	1.71 **2.17**	25
26	4.22 **7.72**	3.37 **5.53**	2.98 **4.64**	2.74 **4.14**	2.59 **3.82**	2.47 **3.59**	2.39 **3.42**	2.32 **3.29**	2.27 **3.17**	2.22 **3.09**	2.18 **3.02**	2.15 **2.96**	2.10 **2.86**	2.05 **2.77**	1.99 **2.66**	1.95 **2.58**	1.90 **2.50**	1.85 **2.41**	1.82 **2.36**	1.78 **2.28**	1.76 **2.25**	1.72 **2.19**	1.70 **2.15**	1.69 **2.13**	26

Table 3 (cont'd)

Degrees of freedom for the denominator	Degrees of freedom for the numerator																								
	1	2	3	4	5	6	7	8	9	10	11	12	14	16	20	24	30	40	50	75	100	200	500	∞	
27	4.21 **7.68**	3.35 **5.49**	2.96 **4.60**	2.73 **4.11**	2.57 **3.79**	2.46 **3.56**	2.37 **3.39**	2.30 **3.26**	2.25 **3.14**	2.20 **3.06**	2.16 **2.98**	2.13 **2.93**	2.08 **2.83**	2.03 **2.74**	1.97 **2.63**	1.93 **2.55**	1.88 **2.47**	1.84 **2.38**	1.80 **2.33**	1.76 **2.25**	1.74 **2.21**	1.71 **2.16**	1.68 **2.12**	1.67 **2.10**	27
28	4.20 **7.64**	3.34 **5.45**	2.95 **4.57**	2.71 **4.07**	2.56 **3.76**	2.44 **3.53**	2.36 **3.36**	2.29 **3.23**	2.24 **3.11**	2.19 **3.03**	2.15 **2.95**	2.12 **2.90**	2.06 **2.80**	2.02 **2.71**	1.96 **2.60**	1.91 **2.52**	1.87 **2.44**	1.81 **2.35**	1.78 **2.30**	1.75 **2.22**	1.72 **2.18**	1.69 **2.13**	1.67 **2.09**	1.65 **2.06**	28
29	4.18 **7.60**	3.33 **5.42**	2.93 **4.54**	2.70 **4.04**	2.54 **3.73**	2.43 **3.50**	2.35 **3.33**	2.28 **3.20**	2.22 **3.08**	2.18 **3.00**	2.14 **2.92**	2.10 **2.87**	2.05 **2.77**	2.00 **2.68**	1.94 **2.57**	1.90 **2.49**	1.85 **2.41**	1.80 **2.32**	1.77 **2.27**	1.73 **2.19**	1.71 **2.15**	1.68 **2.10**	1.65 **2.06**	1.64 **2.03**	29
30	4.17 **7.56**	3.32 **5.39**	2.92 **4.51**	2.69 **4.02**	2.53 **3.70**	2.42 **3.47**	2.34 **3.30**	2.27 **3.17**	2.21 **3.06**	2.16 **2.98**	2.12 **2.90**	2.09 **2.84**	2.04 **2.74**	1.99 **2.66**	1.93 **2.55**	1.89 **2.47**	1.84 **2.38**	1.79 **2.29**	1.76 **2.24**	1.72 **2.16**	1.69 **2.13**	1.66 **2.07**	1.64 **2.03**	1.62 **2.01**	30
32	4.15 **7.50**	3.30 **5.34**	2.90 **4.46**	2.67 **3.97**	2.51 **3.66**	2.40 **3.42**	2.32 **3.25**	2.25 **3.12**	2.19 **3.01**	2.14 **2.94**	2.10 **2.86**	2.07 **2.80**	2.02 **2.70**	1.97 **2.62**	1.91 **2.51**	1.86 **2.42**	1.82 **2.34**	1.76 **2.25**	1.74 **2.20**	1.69 **2.12**	1.67 **2.08**	1.64 **2.02**	1.61 **1.98**	1.59 **1.96**	32
34	4.13 **7.44**	3.28 **5.29**	2.88 **4.42**	2.65 **3.93**	2.49 **3.61**	2.38 **3.38**	2.30 **3.21**	2.23 **3.08**	2.17 **2.97**	2.12 **2.89**	2.08 **2.82**	2.05 **2.76**	2.00 **2.66**	1.95 **2.58**	1.89 **2.47**	1.84 **2.38**	1.80 **2.30**	1.74 **2.21**	1.71 **2.15**	1.67 **2.08**	1.64 **2.04**	1.61 **1.98**	1.59 **1.94**	1.57 **1.91**	34
36	4.11 **7.39**	3.26 **5.25**	2.86 **4.38**	2.63 **3.89**	2.48 **3.58**	2.36 **3.35**	2.28 **3.18**	2.21 **3.04**	2.15 **2.94**	2.10 **2.86**	2.06 **2.78**	2.03 **2.72**	1.98 **2.62**	1.93 **2.54**	1.87 **2.43**	1.82 **2.35**	1.78 **2.26**	1.72 **2.17**	1.69 **2.12**	1.65 **2.04**	1.62 **2.00**	1.59 **1.94**	1.56 **1.90**	1.55 **1.87**	36
38	4.10 **7.35**	3.25 **5.21**	2.85 **4.34**	2.62 **3.86**	2.46 **3.54**	2.35 **3.32**	2.26 **3.15**	2.19 **3.02**	2.14 **2.91**	2.09 **2.82**	2.05 **2.75**	2.02 **2.69**	1.96 **2.59**	1.92 **2.51**	1.85 **2.40**	1.80 **2.32**	1.76 **2.22**	1.71 **2.14**	1.67 **2.08**	1.63 **2.00**	1.60 **1.97**	1.57 **1.90**	1.54 **1.86**	1.53 **1.84**	38
40	4.08 **7.31**	3.23 **5.18**	2.84 **4.31**	2.61 **3.83**	2.45 **3.51**	2.34 **3.29**	2.25 **3.12**	2.18 **2.99**	2.12 **2.88**	2.07 **2.80**	2.04 **2.73**	2.00 **2.66**	1.95 **2.56**	1.90 **2.49**	1.84 **2.37**	1.79 **2.29**	1.74 **2.20**	1.69 **2.11**	1.66 **2.05**	1.61 **1.97**	1.59 **1.94**	1.55 **1.88**	1.53 **1.84**	1.51 **1.81**	40
42	4.07 **7.27**	3.22 **5.15**	2.83 **4.29**	2.59 **3.80**	2.44 **3.49**	2,32 **3.26**	2.24 **3.10**	2.17 **2.96**	2.11 **2.86**	2.06 **2.77**	2.02 **2.70**	1.99 **2.64**	1.94 **2.54**	1.89 **2.46**	1.82 **2.35**	1.78 **2.26**	1.73 **2.17**	1.68 **2.08**	1.64 **2.02**	1.60 **1.94**	1.57 **1.91**	1.54 **1.85**	1.51 **1.80**	1.49 **1.78**	42
44	4.06 **7.24**	3.21 **5.12**	2.82 **4.26**	2.58 **3.78**	2.43 **3.46**	2.31 **3.24**	2.23 **3.07**	2.16 **2.94**	2.10 **2.84**	2.05 **2.75**	2.01 **2.68**	1.98 **2.62**	1.92 **2.52**	1.88 **2.44**	1.81 **2.32**	1.76 **2.24**	1.72 **2.15**	1.66 **2.06**	1.63 **2.00**	1.58 **1.92**	1.56 **1.88**	1.52 **1.82**	1.50 **1.78**	1.48 **1.75**	44
46	4.05 **7.21**	3.20 **5.10**	2.81 **4.24**	2.57 **3.76**	2.42 **3.44**	2.30 **3.22**	2.22 **3.05**	2.14 **2.92**	2.09 **2.82**	2.04 **2.73**	2.00 **2.66**	1.97 **2.60**	1.91 **2.50**	1.87 **2.42**	1.80 **2.30**	1.75 **2.22**	1.71 **2.13**	1.65 **2.04**	1.62 **1.98**	1.57 **1.90**	1.54 **1.86**	1.51 **1.80**	1.48 **1.76**	1.46 **1.72**	46
48	4.04 **7.19**	3.19 **5.08**	2.80 **4.22**	2.56 **3.74**	2.41 **3.42**	2.30 **3.20**	2.21 **3.04**	2.14 **2.90**	2.08 **2.80**	2.03 **2.71**	1.99 **2.64**	1.96 **2.58**	1.90 **2.48**	1.86 **2.40**	1.79 **2.28**	1.74 **2.20**	1.70 **2.11**	1.64 **2.02**	1.61 **1.96**	1.56 **1.88**	1.53 **1.84**	1.50 **1.78**	1.47 **1.73**	1.45 **1.70**	48

Table 3 (cont'd)

Degrees of freedom for the denominator	Degrees of freedom for the numerator: 1	2	3	4	5	6	7	8	9	10	11	12	14	16	20	24	30	40	50	75	100	200	500	∞	
50	4.03	3.18	2.79	2.56	2.40	2.29	2.20	2.13	2.07	2.02	1.98	1.95	1.90	1.85	1.78	1.74	1.69	1.63	1.60	1.55	1.52	1.48	1.46	1.44	50
	7.17	**5.06**	**4.20**	**3.72**	**3.41**	**3.18**	**3.02**	**2.88**	**2.78**	**2.70**	**2.62**	**2.56**	**2.46**	**2.39**	**2.26**	**2.18**	**2.10**	**2.00**	**1.94**	**1.86**	**1.82**	**1.76**	**1.71**	**1.68**	
55	4.02	3.17	2.78	2.54	2.38	2.27	2.18	2.11	2.05	2.00	1.97	1.93	1.88	1.83	1.76	1.72	1.67	1.61	1.58	1.52	1.50	1.46	1.43	1.41	55
	7.12	**5.01**	**4.16**	**3.68**	**3.37**	**3.15**	**2.98**	**2.85**	**2.75**	**2.66**	**2.59**	**2.53**	**2.43**	**2.35**	**2.23**	**2.15**	**2.06**	**1.96**	**1.90**	**1.82**	**1.78**	**1.71**	**1.66**	**1.64**	
60	4.00	3.15	2.76	2.52	2.37	2.25	2.17	2.10	2.04	1.99	1.95	1.92	1.86	1.81	1.75	1.70	1.65	1.59	1.56	1.50	1.48	1.44	1.41	1.39	60
	7.08	**4.98**	**4.13**	**3.65**	**3.34**	**3.12**	**2.95**	**2.82**	**2.72**	**2.63**	**2.56**	**2.50**	**2.40**	**2.32**	**2.20**	**2.12**	**2.03**	**1.93**	**1.87**	**1.79**	**1.74**	**1.68**	**1.63**	**1.60**	
65	3.99	3.14	2.75	2.51	2.36	2.24	2.15	2.08	2.02	1.98	1.94	1.90	1.85	1.80	1.73	1.68	1.63	1.57	1.54	1.49	1.46	1.42	1.39	1.37	65
	7.04	**4.95**	**4.10**	**3.62**	**3.31**	**3.09**	**2.93**	**2.79**	**2.70**	**2.61**	**2.54**	**2.47**	**2.37**	**2.30**	**2.18**	**2.09**	**2.00**	**1.90**	**1.84**	**1.76**	**1.71**	**1.64**	**1.60**	**1.56**	
70	3.98	3.13	2.74	2.50	2.35	2.23	2.14	2.07	2.01	1.97	1.93	1.89	1.84	1.79	1.72	1.67	1.62	1.56	1.53	1.47	1.45	1.40	1.37	1.35	70
	7.01	**4.92**	**4.08**	**3.60**	**3.29**	**3.07**	**2.91**	**2.77**	**2.67**	**2.59**	**2.51**	**2.45**	**2.35**	**2.28**	**2.15**	**2.07**	**1.98**	**1.88**	**1.82**	**1.74**	**1.69**	**1.62**	**1.56**	**1.53**	
80	3.96	3.11	2.72	2.48	2.33	2.21	2.12	2.05	1.99	1.95	1.91	1.88	1.82	1.77	1.70	1.65	1.60	1.54	1.51	1.45	1.42	1.38	1.35	1.32	80
	6.96	**4.88**	**4.04**	**3.56**	**3.25**	**3.04**	**2.87**	**2.74**	**2.64**	**2.55**	**2.48**	**2.41**	**2.32**	**2.24**	**2.11**	**2.03**	**1.94**	**1.84**	**1.78**	**1.70**	**1.65**	**1.57**	**1.52**	**1.49**	
100	3.94	3.09	2.70	2.46	2.30	2.19	2.10	2.03	1.97	1.92	1.88	1.85	1.79	1.75	1.68	1.63	1.57	1.51	1.48	1.42	1.39	1.34	1.30	1.28	100
	6.90	**4.82**	**3.98**	**3.51**	**3.20**	**2.99**	**2.82**	**2.69**	**2.59**	**2.51**	**2.43**	**2.36**	**2.26**	**2.19**	**2.06**	**1.98**	**1.89**	**1.79**	**1.73**	**1.64**	**1.59**	**1.51**	**1.46**	**1.43**	
125	3.92	3.07	2.68	2.44	2.29	2.17	2.08	2.01	1.95	1.90	1.86	1.83	1.77	1.72	1.65	1.60	1.55	1.49	1.45	1.39	1.36	1.31	1.27	1.25	125
	6.84	**4.78**	**3.94**	**3.47**	**3.17**	**2.95**	**2.79**	**2.65**	**2.56**	**2.47**	**2.40**	**2.33**	**2.23**	**2.15**	**2.03**	**1.94**	**1.85**	**1.75**	**1.68**	**1.59**	**1.54**	**1.46**	**1.40**	**1.37**	
150	3.91	3.06	2.67	2.43	2.27	2.16	2.07	2.00	1.94	1.89	1.85	1.82	1.76	1.71	1.64	1.59	1.54	1.47	1.44	1.37	1.34	1.29	1.25	1.22	150
	6.81	**4.75**	**3.91**	**3.44**	**3.14**	**2.92**	**2.76**	**2.62**	**2.53**	**2.44**	**2.37**	**2.30**	**2.20**	**2.12**	**2.00**	**1.91**	**1.83**	**1.72**	**1.66**	**1.56**	**1.51**	**1.43**	**1.37**	**1.33**	
200	3.89	3.04	2.65	2.41	2.26	2.14	2.05	1.98	1.92	1.87	1.83	1.80	1.74	1.69	1.62	1.57	1.52	1.45	1.42	1.35	1.32	1.26	1.22	1.19	200
	6.76	**4.71**	**3.88**	**3.41**	**3.11**	**2.90**	**2.73**	**2.60**	**2.50**	**2.41**	**2.34**	**2.28**	**2.17**	**2.09**	**1.97**	**1.88**	**1.79**	**1.69**	**1.62**	**1.53**	**1.48**	**1.39**	**1.33**	**1.28**	
400	3.86	3.02	2.62	2.39	2.23	2.12	2.03	1.96	1.90	1.85	1.81	1.78	1.72	1.67	1.60	1.54	1.49	1.42	1.38	1.32	1.28	1.22	1.16	1.13	400
	6.70	**4.66**	**3.83**	**3.36**	**3.06**	**2.85**	**2.69**	**2.55**	**2.46**	**2.37**	**2.29**	**2.23**	**2.12**	**2.04**	**1.92**	**1.84**	**1.74**	**1.64**	**1.57**	**1.47**	**1.42**	**1.32**	**1.24**	**1.19**	
1000	3.85	3.00	2.61	2.38	2.22	2.10	2.02	1.95	1.89	1.84	1.80	1.76	1.70	1.65	1.58	1.53	1.47	1.41	1.36	1.30	1.26	1.19	1.13	1.08	1000
	6.66	**4.62**	**3.80**	**3.34**	**3.04**	**2.82**	**2.66**	**2.53**	**2.43**	**2.34**	**2.26**	**2.20**	**2.09**	**2.01**	**1.89**	**1.81**	**1.71**	**1.51**	**1.54**	**1.44**	**1.38**	**1.28**	**1.19**	**1.11**	
∞	3.84	2.99	2.60	2.37	2.21	2.09	2.01	1.94	1.88	1.83	1.79	1.75	1.69	1.64	1.57	1.52	1.46	1.40	1.35	1.28	1.24	1.17	1.11	1.00	∞
	6.64	**4.60**	**3.78**	**3.32**	**3.02**	**2.80**	**2.64**	**2.51**	**2.41**	**2.32**	**2.24**	**2.18**	**2.07**	**1.99**	**1.87**	**1.79**	**1.69**	**1.59**	**1.52**	**1.41**	**1.36**	**1.25**	**1.15**	**1.00**	

Source: From *Fundamental Statistics for Psychology,* copyright © 1975 by Robert McCall, reproduced by permission of Harcourt Brace Jovanovich, Inc.

Table 4 Critical Values of Chi-Square

df	Level of significance for a non-directional test					
	.20	.10	.05	.02	.01	.001
1	1.64	2.71	3.84	5.41	6.64	10.83
2	3.22	4.60	5.99	7.82	9.21	13.82
3	4.64	6.25	7.82	9.84	11.34	16.27
4	5.99	7.78	9.49	11.67	13.28	18.46
5	7.29	9.24	11.07	13.39	15.09	20.52
6	8.56	10.64	12.59	15.03	16.81	22.46
7	9.80	12.02	14.07	16.62	18.48	24.32
8	11.03	13.36	15.51	18.17	20.09	26.12
9	12.24	14.68	16.92	19.68	21.67	27.88
10	13.44	15.99	18.31	21.16	23.21	29.59
11	14.63	17.28	19.68	22.62	24.72	31.26
12	15.81	18.55	21.03	24.05	26.22	32.91
13	16.98	19.81	22.36	25.47	27.69	34.53
14	18.15	21.06	23.68	26.87	29.14	36.12
15	19.31	22.31	25.00	28.26	30.58	37.70
16	20.46	23.54	26.30	29.63	32.00	39.29
17	21.62	24.77	27.59	31.00	33.41	40.75
18	22.76	25.99	28.87	32.35	34.80	42.31
19	23.90	27.20	30.14	33.69	36.19	43.82
20	25.04	28.41	31.41	35.02	37.57	45.32
21	26.17	29.62	32.67	36.34	38.93	46.80
22	27.30	30.81	33.92	37.66	40.29	48.27
23	28.43	32.01	35.17	38.97	41.64	49.73
24	29.55	33.20	36.42	40.27	42.98	51.18
25	30.68	34.38	37.65	41.57	44.31	52.62
26	31.80	35.56	38.88	42.86	45.64	54.05
27	32.91	36.74	40.11	44.14	46.96	55.48
28	34.03	37.92	41.34	45.42	48.28	56.89
29	35.14	39.09	42.69	46.69	49.59	58.30
30	36.25	40.26	43.77	47.96	50.89	59.70
32	38.47	42.59	46.19	50.49	53.49	62.49
34	40.68	44.90	48.60	53.00	56.06	65.25
36	42.88	47.21	51.00	55.49	58.62	67.99
38	45.08	49.51	53.38	57.97	61.16	70.70
40	47.27	51.81	55.76	60.44	63.69	73.40
44	51.64	56.37	60.48	65.34	68.71	78.75
48	55.99	60.91	65.17	70.20	73.68	84.04
52	60.33	65.42	69.83	75.02	78.62	89.27
56	64.66	69.92	74.47	79.82	83.51	94.46
60	68.97	74.40	79.08	84.58	88.38	99.61

Source: From *Fundamental Statistics for Psychology,* copyright © 1975 by Robert McCall, reproduced by permission of Harcourt Brace Jovanovich, Inc.

Table 5 Studentized Range Points for Tukey Test

Error df	α	r = number of means or number of steps between ordered means									
		2	3	4	5	6	7	8	9	10	11
5	.05	3.64	4.60	5.22	5.67	6.03	6.33	6.58	6.80	6.99	7.17
	.01	5.70	6.98	7.80	8.42	8.91	9.32	9.67	9.97	10.24	10.48
6	.05	3.46	4.34	4.90	5.30	5.63	5.90	6.12	6.32	6.49	6.65
	.01	5.24	6.33	7.03	7.56	7.97	8.32	8.61	8.87	9.10	9.30
7	.05	3.34	4.16	4.68	5.06	5.36	5.61	5.82	6.00	6.16	6.30
	.01	4.95	5.92	6.54	7.01	7.37	7.68	7.94	8.17	8.37	8.55
8	.05	3.26	4.04	4.53	4.89	5.17	5.40	5.60	5.77	5.92	6.05
	.01	4.75	5.64	6.20	6.62	6.96	7.24	7.47	7.68	7.86	8.03
9	.05	3.20	3.95	4.41	4.76	5.02	5.24	5.43	5.59	5.74	5.87
	.01	4.60	5.43	5.96	6.35	6.66	6.91	7.13	7.33	7.49	7.65
10	.05	3.15	3.88	4.33	4.65	4.91	5.12	5.30	5.46	5.60	5.72
	.01	4.48	5.27	5.77	6.14	6.43	6.67	6.87	7.05	7.21	7.36
11	.05	3.11	3.82	4.26	4.57	4.82	5.03	5.20	5.35	5.49	5.61
	.01	4.39	5.15	5.62	5.97	6.25	6.48	6.67	6.84	6.99	7.13
12	.05	3.08	3.77	4.20	4.51	4.75	4.95	5.12	5.27	5.39	5.51
	.01	4.32	5.05	5.50	5.84	6.10	6.32	6.51	6.67	6.81	6.94
13	.05	3.06	3.73	4.15	4.45	4.69	4.88	5.05	5.19	5.32	5.43
	.01	4.26	4.96	5.40	5.73	5.98	6.19	6.37	6.53	6.67	6.79
14	.05	3.03	3.70	4.11	4.41	4.64	4.83	4.99	5.13	5.25	5.36
	.01	4.21	4.89	5.32	5.63	5.88	6.08	6.26	6.41	6.54	6.66
15	.05	3.01	3.67	4.08	4.37	4.59	4.78	4.94	5.08	5.20	5.31
	.01	4.17	4.84	5.25	5.56	5.80	5.99	6.16	6.31	6.44	6.55
16	.05	3.00	3.65	4.05	4.33	4.56	4.74	4.90	5.03	5.15	5.26
	.01	4.13	4.79	5.19	5.49	5.72	5.92	6.08	6.22	6.35	6.46
17	.05	2.98	3.63	4.02	4.30	4.52	4.70	4.86	4.99	5.11	5.21
	.01	4.10	4.74	5.14	5.43	5.66	5.85	6.01	6.15	6.27	6.38
18	.05	2.97	3.61	4.00	4.28	4.49	4.67	4.82	4.96	5.07	5.17
	.01	4.07	4.70	5.09	5.38	5.60	5.79	5.94	6.08	6.20	6.31
19	.05	2.96	3.59	3.98	4.25	4.47	4.65	4.79	4.92	5.04	5.14
	.01	4.05	4.67	5.05	5.33	5.55	5.73	5.89	6.02	6.14	6.25
20	.05	2.95	3.58	3.96	4.23	4.45	4.62	4.77	4.90	5.01	5.11
	.01	4.02	4.64	5.02	5.29	5.51	5.69	5.84	5.97	6.09	6.19
24	.05	2.92	3.53	3.90	4.17	4.37	4.54	4.68	4.81	4.92	5.01
	.01	3.96	4.55	4.91	5.17	5.37	5.54	5.69	5.81	5.92	6.02
30	.05	2.89	3.49	3.85	4.10	4.30	4.46	4.60	4.72	4.82	4.92
	.01	3.89	4.45	4.80	5.05	5.24	5.40	5.54	5.65	5.76	5.85
40	.05	2.86	3.44	3.79	4.04	4.23	4.39	4.52	4.63	4.73	4.82
	.01	3.82	4.37	4.70	4.93	5.11	5.26	5.39	5.50	5.60	5.69
60	.05	2.83	3.40	3.74	3.98	4.16	4.31	4.44	4.55	4.65	4.73
	.01	3.76	4.28	4.59	4.82	4.99	5.13	5.25	5.36	5.45	5.53
120	.05	2.80	3.36	3.68	3.92	4.10	4.24	4.36	4.47	4.56	4.64
	.01	3.70	4.20	4.50	4.71	4.87	5.01	5.12	5.21	5.30	5.37
∞	.05	2.77	3.31	3.63	3.86	4.03	4.17	4.29	4.39	4.47	4.55
	.01	3.64	4.12	4.40	4.60	4.76	4.88	4.99	5.08	5.16	5.23

Table 5 (cont'd)

r = number of means or number of steps between ordered means										
12	13	14	15	16	17	18	19	20	α	Error df
7.32	7.47	7.60	7.72	7.83	7.93	8.03	8.12	8.21	.05	5
10.70	10.89	11.08	11.24	11.40	11.55	11.68	11.81	11.93	.01	
6.79	6.92	7.03	7.14	7.24	7.34	7.43	7.51	7.59	.05	6
9.48	9.65	9.81	9.95	10.08	10.21	10.32	10.43	10.54	.01	
6.43	6.55	6.66	6.76	6.85	6.94	7.02	7.10	7.17	.05	7
8.71	8.86	9.00	9.12	9.24	9.35	9.46	9.55	9.65	.01	
6.18	6.29	6.39	6.48	6.57	6.65	6.73	6.80	6.87	.05	8
8.18	8.31	8.44	8.55	8.66	8.76	8.85	8.94	9.03	.01	
5.98	6.09	6.19	6.28	6.36	6.44	6.51	6.58	6.64	.05	9
7.78	7.91	8.03	8.13	8.23	8.33	8.41	8.49	8.57	.01	
5.83	5.93	6.03	6.11	6.19	6.27	6.34	6.40	6.47	.05	10
7.49	7.60	7.71	7.81	7.91	7.99	8.08	8.15	8.23	.01	
5.71	5.81	5.90	5.98	6.06	6.13	6.20	6.27	6.33	.05	11
7.25	7.36	7.46	7.56	7.65	7.73	7.81	7.88	7.95	.01	
5.61	5.71	5.80	5.88	5.95	6.02	6.09	6.15	6.21	.05	12
7.06	7.17	7.26	7.36	7.44	7.52	7.59	7.66	7.73	.01	
5.53	5.63	5.71	5.79	5.86	5.93	5.99	6.05	6.11	.05	13
6.90	7.01	7.10	7.19	7.27	7.35	7.42	7.48	7.55	.01	
5.46	5.55	5.64	5.71	5.79	5.85	5.91	5.97	6.03	.05	14
6.77	6.87	6.96	7.05	7.13	7.20	7.27	7.33	7.39	.01	
5.40	5.49	5.57	5.65	5.72	5.78	5.85	5.90	5.96	.05	15
6.66	6.76	6.84	6.93	7.00	7.07	7.14	7.20	7.26	.01	
5.35	5.44	5.52	5.59	5.66	5.73	5.79	5.84	5.90	.05	16
6.56	6.66	6.74	6.82	6.90	6.97	7.03	7.09	7.15	.01	
5.31	5.39	5.47	5.54	5.61	5.67	5.73	5.79	5.84	.05	17
6.48	6.57	6.66	6.73	6.81	6.87	6.94	7.00	7.05	.01	
5.27	5.35	5.43	5.50	5.57	5.63	5.69	5.74	5.79	.05	18
6.41	6.50	6.58	6.65	6.73	6.79	6.85	6.91	6.97	.01	
5.23	5.31	5.39	5.46	5.53	5.59	5.65	5.70	5.75	.05	19
6.34	6.43	6.51	6.58	6.65	6.72	6.78	6.84	6.89	.01	
5.20	5.28	5.36	5.43	5.49	5.55	5.61	5.66	5.71	.05	20
6.28	6.37	6.45	6.52	6.59	6.65	6.71	6.77	6.82	.01	
5.10	5.18	5.25	5.32	5.38	5.44	5.49	5.55	5.59	.05	24
6.11	6.19	6.26	6.33	6.39	6.45	6.51	6.56	6.61	.01	
5.00	5.08	5.15	5.21	5.27	5.33	5.38	5.43	5.47	.05	30
5.93	6.01	6.08	6.14	6.20	6.26	6.31	6.36	6.41	.01	
4.90	4.98	5.04	5.11	5.16	5.22	5.27	5.31	5.36	.05	40
5.76	5.83	5.90	5.96	6.02	6.07	6.12	6.16	6.21	.01	
4.81	4.88	4.94	5.00	5.06	5.11	5.15	5.20	5.24	.05	60
5.60	5.67	5.73	5.78	5.84	5.89	5.93	5.97	6.01	.01	
4.71	4.78	4.84	4.90	4.95	5.00	5.04	5.09	5.13	.05	120
5.44	5.50	5.56	5.61	5.66	5.71	5.75	5.79	5.83	.01	
4.62	4.68	4.74	4.80	4.85	4.89	4.93	4.97	5.01	.05	∞
5.29	5.35	5.40	5.45	5.49	5.54	5.57	5.61	5.65	.01	

Source: Biometrika Trustees, *Biometrika Tables for Statisticians,* Vol. 1, 3rd ed. 1966. Reprinted with permission of the Biometrika Trustees.

Table 6 Critical Values of Mann-Whitney U

Critical Values of the Mann–Whitney *U*. For a one-tailed test at $\alpha = 0.01$ (roman type) and $\alpha = 0.005$ (boldface type) and for a two-tailed test at $\alpha = 0.02$ (roman type) and $\alpha = 0.01$ (boldface type).

n_2 \ n_1	*1*	*2*	*3*	*4*	*5*	*6*	*7*	*8*	*9*	*10*	*11*	*12*	*13*	*14*	*15*	*16*	*17*	*18*	*19*	*20*
1	—[b]	—	—	—	—	—	—	—	—	—	—	—	—	—	—	—	—	—	—	—
2	—	—	—	—	—	—	—	—	—	—	—	—	0	0	0	0	0	0	1	1
													—	—	—	—	—	—	**0**	**0**
3	—	—	—	—	—	—	0	0	1	1	1	2	2	2	3	3	4	4	4	5
							—	—	**0**	**0**	**0**	**1**	**1**	**1**	**2**	**2**	**2**	**2**	**3**	**3**
4	—	—	—	—	0	1	1	2	3	3	4	5	5	6	7	7	8	9	9	10
					—	**0**	**0**	**1**	**1**	**2**	**2**	**3**	**3**	**4**	**5**	**5**	**6**	**6**	**7**	**8**
5	—	—	—	0	1	2	3	4	5	6	7	8	9	10	11	12	13	14	15	16
				—	**0**	**1**	**1**	**2**	**3**	**4**	**5**	**6**	**7**	**7**	**8**	**9**	**10**	**11**	**12**	**13**
6	—	—	—	1	2	3	4	6	7	8	9	11	12	13	15	16	18	19	20	22
				0	**1**	**2**	**3**	**4**	**5**	**6**	**7**	**9**	**10**	**11**	**12**	**13**	**15**	**16**	**17**	**18**
7	—	—	0	1	3	4	6	7	9	11	12	14	16	17	19	21	23	24	26	28
			—	**0**	**1**	**3**	**4**	**6**	**7**	**9**	**10**	**12**	**13**	**15**	**16**	**18**	**19**	**21**	**22**	**24**
8	—	—	0	2	4	6	7	9	11	13	15	17	20	22	24	26	28	30	32	34
			—	**1**	**2**	**4**	**6**	**7**	**9**	**11**	**13**	**15**	**17**	**18**	**20**	**22**	**24**	**26**	**28**	**30**
9	—	—	1	3	5	7	9	11	14	16	18	21	23	26	28	31	33	36	38	40
			0	**1**	**3**	**5**	**7**	**9**	**11**	**13**	**16**	**18**	**20**	**22**	**24**	**27**	**29**	**31**	**33**	**36**
10	—	—	1	3	6	8	11	13	16	19	22	24	27	30	33	36	38	41	44	47
			0	**2**	**4**	**6**	**9**	**11**	**13**	**16**	**18**	**21**	**24**	**26**	**29**	**31**	**34**	**37**	**39**	**42**
11	—	—	1	4	7	9	12	15	18	22	25	28	31	34	37	41	44	47	50	53
			0	**2**	**5**	**7**	**10**	**13**	**16**	**18**	**21**	**24**	**27**	**30**	**33**	**36**	**39**	**42**	**45**	**48**
12	—	—	2	5	8	11	14	17	21	24	28	31	35	38	42	46	49	53	56	60
			1	**3**	**6**	**9**	**12**	**15**	**18**	**21**	**24**	**27**	**31**	**34**	**37**	**41**	**44**	**47**	**51**	**54**
13	—	0	2	5	9	12	16	20	23	27	31	35	39	43	47	51	55	59	63	67
		—	**1**	**3**	**7**	**10**	**13**	**17**	**20**	**24**	**27**	**31**	**34**	**38**	**42**	**45**	**49**	**53**	**56**	**60**
14	—	0	2	6	10	13	17	22	26	30	34	38	43	47	51	56	60	65	69	73
		—	**1**	**4**	**7**	**11**	**15**	**18**	**22**	**26**	**30**	**34**	**38**	**42**	**46**	**50**	**54**	**58**	**63**	**67**
15	—	0	3	7	11	15	19	24	28	33	37	42	47	51	56	61	66	70	75	80
		—	**2**	**5**	**8**	**12**	**16**	**20**	**24**	**29**	**33**	**37**	**42**	**46**	**51**	**55**	**60**	**64**	**69**	**73**
16	—	0	3	7	12	16	21	26	31	36	41	46	51	56	61	66	71	76	82	87
		—	**2**	**5**	**9**	**13**	**18**	**22**	**27**	**31**	**36**	**41**	**45**	**50**	**55**	**60**	**65**	**70**	**74**	**79**
17	—	0	4	8	13	18	23	28	33	38	44	49	55	60	66	71	77	82	88	93
		—	**2**	**6**	**10**	**15**	**19**	**24**	**29**	**34**	**39**	**44**	**49**	**54**	**60**	**65**	**70**	**75**	**81**	**86**
18	—	0	4	9	14	19	24	30	36	41	47	53	59	65	70	76	82	88	94	100
		—	**2**	**6**	**11**	**16**	**21**	**26**	**31**	**37**	**42**	**47**	**53**	**58**	**64**	**70**	**75**	**81**	**87**	**92**
19	—	1	4	9	15	20	26	32	38	44	50	56	63	69	75	82	88	94	101	107
		0	**3**	**7**	**12**	**17**	**22**	**28**	**33**	**39**	**45**	**51**	**56**	**63**	**69**	**74**	**81**	**87**	**93**	**99**
20	—	1	5	10	16	22	28	34	40	47	53	60	67	73	80	87	93	100	107	114
		0	**3**	**8**	**13**	**18**	**24**	**30**	**36**	**42**	**48**	**54**	**60**	**67**	**73**	**79**	**86**	**92**	**99**	**105**

Table 6 (cont'd)

Critical values for a one-tailed test at $\alpha = 0.05$ (roman type) and $\alpha = 0.025$ (boldface type) and for a two-tailed test at $\alpha = 0.10$ (roman type) and $\alpha = 0.05$ (boldface type).

n_2 \ n_1	1	2	3	4	5	6	7	8	9	10	11	12	13	14	15	16	17	18	19	20
1	—	—	—	—	—	—	—	—	—	—	—	—	—	—	—	—	—	—	0	0
																			—	**—**
2	—	—	—	—	0	0	0	1	1	1	1	2	2	2	3	3	3	4	4	4
					—	**—**	**—**	**0**	**0**	**0**	**0**	**1**	**1**	**1**	**1**	**1**	**2**	**2**	**2**	**2**
3	—	—	0	0	1	2	2	3	3	4	5	5	6	7	7	8	9	9	10	11
	—	**—**	**—**	**—**	**0**	**1**	**1**	**2**	**2**	**3**	**3**	**4**	**4**	**5**	**5**	**6**	**6**	**7**	**7**	**8**
4	—	—	0	1	2	3	4	5	6	7	8	9	10	11	12	14	15	16	17	18
	—	**—**	**—**	**0**	**1**	**2**	**3**	**4**	**4**	**5**	**6**	**7**	**8**	**9**	**10**	**11**	**11**	**12**	**13**	**13**
5	—	0	1	2	4	5	6	8	9	11	12	13	15	16	18	19	20	22	23	25
	—	**—**	**0**	**1**	**2**	**3**	**5**	**6**	**7**	**8**	**9**	**11**	**12**	**13**	**14**	**15**	**17**	**18**	**19**	**20**
6	—	0	2	3	5	7	8	10	12	14	16	17	19	21	23	25	26	28	30	32
	—	**—**	**1**	**2**	**3**	**5**	**6**	**8**	**10**	**11**	**13**	**14**	**16**	**17**	**19**	**21**	**22**	**24**	**25**	**27**
7	—	0	2	4	6	8	11	13	15	17	19	21	24	26	28	30	33	35	37	39
	—	**—**	**1**	**3**	**5**	**6**	**8**	**10**	**12**	**14**	**16**	**18**	**20**	**22**	**24**	**26**	**28**	**30**	**32**	**34**
8	—	1	3	5	8	10	13	15	18	20	23	26	28	31	33	36	39	41	44	47
	—	**0**	**2**	**4**	**6**	**8**	**10**	**13**	**15**	**17**	**19**	**22**	**24**	**26**	**29**	**31**	**34**	**36**	**38**	**41**
9	—	1	3	6	9	12	15	18	21	24	27	30	33	36	39	42	45	48	51	54
	—	**0**	**2**	**4**	**7**	**10**	**12**	**15**	**17**	**20**	**23**	**26**	**28**	**31**	**34**	**37**	**39**	**42**	**45**	**48**
10	—	1	4	7	11	14	17	20	24	27	31	34	37	41	44	48	51	55	58	62
	—	**0**	**3**	**5**	**8**	**11**	**14**	**17**	**20**	**23**	**26**	**29**	**33**	**36**	**39**	**42**	**45**	**48**	**52**	**55**
11	—	1	5	8	12	16	19	23	27	31	34	38	42	46	50	54	57	61	65	69
	—	**0**	**3**	**6**	**9**	**13**	**16**	**19**	**23**	**26**	**30**	**33**	**37**	**40**	**44**	**47**	**51**	**55**	**58**	**62**
12	—	2	5	9	13	17	21	26	30	34	38	42	47	51	55	60	64	68	72	77
	—	**1**	**4**	**7**	**11**	**14**	**18**	**22**	**26**	**29**	**33**	**37**	**41**	**45**	**49**	**53**	**57**	**61**	**65**	**69**
13	—	2	6	10	15	19	24	28	33	37	42	47	51	56	61	65	70	75	80	84
	—	**1**	**4**	**8**	**12**	**16**	**20**	**24**	**28**	**33**	**37**	**41**	**45**	**50**	**54**	**59**	**63**	**67**	**72**	**76**
14	—	2	7	11	16	21	26	31	36	41	46	51	56	61	66	71	77	82	87	92
	—	**1**	**5**	**9**	**13**	**17**	**22**	**26**	**31**	**36**	**40**	**45**	**50**	**55**	**59**	**64**	**67**	**74**	**78**	**83**
15	—	3	7	12	18	23	28	33	39	44	50	55	61	66	72	77	83	88	94	100
	—	**1**	**5**	**10**	**14**	**19**	**24**	**29**	**34**	**39**	**44**	**49**	**54**	**59**	**64**	**70**	**75**	**80**	**85**	**90**
16	—	3	8	14	19	25	30	36	42	48	54	60	65	71	77	83	89	95	101	107
	—	**1**	**6**	**11**	**15**	**21**	**26**	**31**	**37**	**42**	**47**	**53**	**59**	**64**	**70**	**75**	**81**	**86**	**92**	**98**
17	—	3	9	15	20	26	33	39	45	51	57	64	70	77	83	89	96	102	109	115
	—	**2**	**6**	**11**	**17**	**22**	**28**	**34**	**39**	**45**	**51**	**57**	**63**	**67**	**75**	**81**	**87**	**93**	**99**	**105**
18	—	4	9	16	22	28	35	41	48	55	61	68	75	82	88	95	102	109	116	123
	—	**2**	**7**	**12**	**18**	**24**	**30**	**36**	**42**	**48**	**55**	**61**	**67**	**74**	**80**	**86**	**93**	**99**	**106**	**112**
19	0	4	10	17	23	30	37	44	51	58	65	72	80	87	94	101	109	116	123	130
	—	**2**	**7**	**13**	**19**	**25**	**32**	**38**	**45**	**52**	**58**	**65**	**72**	**78**	**85**	**92**	**99**	**106**	**113**	**119**
20	0	4	11	18	25	32	39	47	54	62	69	77	84	92	100	107	115	123	130	138
	—	**2**	**8**	**13**	**20**	**27**	**34**	**41**	**48**	**55**	**62**	**69**	**76**	**83**	**90**	**98**	**105**	**112**	**119**	**127**

Source: Kirk, *Introductory Statistics,* (Monterey, Cal.: Brooks/Cole, 1978).

Table 7 Critical Values of Wilcoxon T

	Level of Significance for a One-Tailed Test			
	0.05	*0.025*	*0.01*	*0.005*
	Level of Significance for a Two-Tailed Test			
n	*0.10*	*0.05*	*0.02*	*0.01*
5	0	—	—	—
6	2	0	—	—
7	3	2	0	—
8	5	3	1	0
9	8	5	3	1
10	10	8	5	3
11	13	10	7	5
12	17	13	9	7
13	21	17	12	9
14	25	21	15	12
15	30	25	19	15
16	35	29	23	19
17	41	34	27	23
18	47	40	32	27
19	53	46	37	32
20	60	52	43	37

Source: Kirk, *Introductory Statistics,* (Monterey, Cal.: Brooks/Cole, 1978).

Table 8 Critical Values of Spearman rho

	Significance level for a directional test at			
	.05	.025	.005	.001
	Significance level for a non-directional test at			
N	.10	.05	.01	.002
5	.900	1.000		
6	.829	.886	1.000	
7	.715	.786	.929	1.000
8	.620	.715	.881	.953
9	.600	.700	.834	.917
10	.564	.649	.794	.879
11	.537	.619	.764	.855
12	.504	.588	.735	.826
13	.484	.561	.704	.797
14	.464	.539	.680	.772
15	.447	.522	.658	.750
16	.430	.503	.636	.730
17	.415	.488	.618	.711
18	.402	.474	.600	.693
19	.392	.460	.585	.676
20	.381	.447	.570	.661
21	.371	.437	.556	.647
22	.361	.426	.544	.633
23	.353	.417	.532	.620
24	.345	.407	.521	.608
25	.337	.399	.511	.597
26	.331	.391	.501	.587
27	.325	.383	.493	.577
28	.319	.376	.484	.567
29	.312	.369	.475	.558
30	.307	.363	.467	.549

Source: From *Fundamental Statistics for Psychology*, copyright © 1975 by Robert McCall, reproduced by permission of Harcourt Brace Jovanovich, Inc.

Index

2
90
25 rings
13 watch
34 blouse
4 gifts
5 X-Shirt
171 Elois' ring
2
342

cash
charge